KNOWLEDGE & DIPLOMACY

Science Advice in the United Nations System

Committee for Survey and Analysis of Science Advice on
Sustainable Development to International Organizations

Development, Security, and Cooperation
Policy and Global Affairs

NATIONAL RESEARCH COUNCIL
OF THE NATIONAL ACADEMIES

The National Academies Press
Washington, D.C.
www.nap.edu

THE NATIONAL ACADEMIES PRESS   500 Fifth Street, N.W.   Washington, D.C. 20001

NOTICE: The project that is the subject of this report was approved by the Governing Board of the National Research Council, whose members are drawn from the councils of the National Academy of Sciences, the National Academy of Engineering, and the Institute of Medicine. The members of the committee responsible for the report were chosen for their special competences and with regard for appropriate balance.

This study was supported by Grant No. B2001-24 between the National Academy of Sciences and the Sloan Foundation. Any opinions, findings, conclusions, or recommendations expressed in this publication are those of the author(s) and do not necessarily reflect the views of the organizations or agencies that provided support for the project.

International Standard Book Number 0-309-08490-3

Copies of this report are available from Policy and Global Affairs, Development, Security and Cooperation Office, The National Academies, 500 5th Street, N.W., Room TN530, Washington, D.C. 20001.

Additional copies of this report are available from the National Academies Press, 500 Fifth Street, N.W., Lockbox 285, Washington, DC 20055; (800) 624-6242 or (202) 334-3313 (in the Washington metropolitan area); Internet, http://www.nap.edu

Copyright 2002 by the National Academy of Sciences. All rights reserved.

Printed in the United States of America

THE NATIONAL ACADEMIES
Advisers to the Nation on Science, Engineering, and Medicine

The **National Academy of Sciences** is a private, nonprofit, self-perpetuating society of distinguished scholars engaged in scientific and engineering research, dedicated to the furtherance of science and technology and to their use for the general welfare. Upon the authority of the charter granted to it by the Congress in 1863, the Academy has a mandate that requires it to advise the federal government on scientific and technical matters. Dr. Bruce M. Alberts is president of the National Academy of Sciences.

The **National Academy of Engineering** was established in 1964, under the charter of the National Academy of Sciences, as a parallel organization of outstanding engineers. It is autonomous in its administration and in the selection of its members, sharing with the National Academy of Sciences the responsibility for advising the federal government. The National Academy of Engineering also sponsors engineering programs aimed at meeting national needs, encourages education and research, and recognizes the superior achievements of engineers. Dr. Wm. A. Wulf is president of the National Academy of Engineering.

The **Institute of Medicine** was established in 1970 by the National Academy of Sciences to secure the services of eminent members of appropriate professions in the examination of policy matters pertaining to the health of the public. The Institute acts under the responsibility given to the National Academy of Sciences by its congressional charter to be an adviser to the federal government and, upon its own initiative, to identify issues of medical care, research, and education. Dr. Harvey V. Fineberg is president of the Institute of Medicine.

The **National Research Council** was organized by the National Academy of Sciences in 1916 to associate the broad community of science and technology with the Academy's purposes of furthering knowledge and advising the federal government. Functioning in accordance with general policies determined by the Academy, the Council has become the principal operating agency of both the National Academy of Sciences and the National Academy of Engineering in providing services to the government, the public, and the scientific and engineering communities. The Council is administered jointly by both Academies and the Institute of Medicine. Dr. Bruce M. Alberts and Dr. Wm. A. Wulf are chair and vice chair, respectively, of the National Research Council.

www.national-academies.org

Committee for Survey and Analysis of Science Advice on Sustainable Development to International Organizations

Robert A. Frosch, Chair (NAE)
Harvard University

Calestous Juma
Harvard University

Phillip M. Smith
McGeary & Smith

Anne G.K. Solomon
Center for Strategic and International Studies

Michael P. Greene, Staff Officer

John Boright, Deputy Executive Director, Policy and Global Affairs

PREFACE

In the international effort to advance human health, welfare, and development while better managing and conserving the environment and natural resources, there is a clear and growing recognition of the role of scientific and technical knowledge in global governance. This has created an urgent need for the United Nations to equip itself with the capability to bring scientific knowledge to inform international decision making. The failure to do so could reduce the ability of the United Nations to continue to be a credible player in international diplomacy. One of the key functions of the United Nations is to alert governments to emerging issues of relevance to international cooperation, including scientific issues. This mandate is codified in Article 99 of the UN Charter, which empowers the Secretary-General to "bring to the attention of the Security Council any matter, which in his opinion, may threaten the maintenance of international peace and security." Carrying out this task requires continuous access to scientific and technical information, which can be provided in the form of "science advice."

The role of science advice for sustainable development was recognized in Agenda 21, the work program of the United Nations Conference on Environment and Development (UNCED). Chapter 31 of Agenda 21 on "Scientific and Technological Communities" specifically called upon States to "strengthen science and technology advice to the highest levels of the United Nations and other international institutions, in order to ensure the inclusion of science and technology know-how in sustainable development policies and strategies." Since the adoption of Agenda 21, there has been a significant increase in awareness of the role of science and technology in sustainable development, adding to the urgency to strengthen the science advice system in the United Nations.

Interest in strengthening the role of science advice for sustainable development builds on a long tradition of bringing the latest available scientific results to bear on international decision making. Over the years, the United Nations has developed a variety of methods for providing science advice to (a) governing bodies of organizations such as conventions and treaties; (b) executive heads and senior management groups; (c) program activities and program development; (d) and member governments. There is considerable variety in the composition and *modus operandi* of the bodies set up to provide science advice. Some activities are carried out through standing bodies (either open to all member countries or limited in size based on rules of traditional practices) or ad hoc groups with mandate and time limitations.

The World Summit on Sustainable Development (WSSD) of September 2002 will reinvigorate the global commitments and achieve a higher level of international solidarity and partnership in the promotion of sustainable development. The United States Department of State, the lead agency for the US at the Summit, has articulated the importance of science-based decision making and is seeking ways to enhance scientific input to the deliberations and the follow-on actions. Accordingly, the Science Advisor of the Secretary of State asked the US National

Academies to survey and analyze the institutional arrangements for science advice to the international agencies involved in international sustainable development, with particular attention to the key fields of water, energy, fisheries, and oceans. The availability of quality science advice to the governing bodies, member governments, and executives of the UN system is critical to the successful achievement of sustainable development. This report, it is hoped, will provide some guidelines for the continuing process of scientific input to critical policy decisions.

The **Committee for Survey and Analysis of Science Advice on Sustainable Development to International Organizations** was asked to compile, interpret, and report available public information that relates to the process for scientific input to the international and multilateral agencies active in the field of sustainable development and to answer the following questions:

- How is scientific information sought and utilized by international, multilateral, and bilateral organizations in the following areas: energy, freshwater quality and use, oceans, and fisheries?

- What is the role of existing scientific bodies, governmental, intergovernmental, and nongovernmental, in providing such information?

- To what extent does the scientific information come from peer-reviewed and independent sources, and how open is the process?

The Committee promptly realized that these questions have no simple answers in the broad universe of United Nations organizations. A consultant examining the problem turned up far more examples that represented science advice in the UN bodies than could possibly be described in a brief report, with few apparent consistent patterns or quality standards. The mechanisms range from overtly political bodies with infrequent, unreported meetings to a panel of independent experts of international stature and an open peer review process. Lacking, however, are standards or principles against which to evaluate the advisory procedures, and a great paucity of information exists for results and outcomes that would have permitted an independent evaluation. The Committee decided that it could make a valuable contribution by examining established science advice mechanisms outside of the UN system in order to extract a set of principles that would be useful for assessing the mechanisms and processes within the system.

This report covers a wide spectrum of United Nations activities, including those activities that fall under the direct purview of the United Nations Secretary-General, organizations that report to the United Nations General Assembly [and whose governing bodies are subsidiaries of the UN General Assembly, such as the United Nations Environment Programme (UNEP) and the United Nations Development Programme (UNDP)], and United Nations system organizations that include specialized agencies with independent governing bodies, such as the United Nations Educational, Scientific, and Cultural Organization (UNESCO), Food and Agriculture Organization of the United Nations (FAO), and World Meteorological Organization (WMO). The wider United Nations system also includes hybrid organizations, as well as programs that have been created through partnerships between different kinds of United Nations organizations but have their own independent governing bodies (such as the Global Environment Facility, which is a partnership between UNDP, UNEP, and the World Bank).

Given the complexity and diversity of United Nations programs, organs, and mandates, this report focuses on the main functions of the United Nations that affect international governance in the fields related to sustainable development, with reference to the taxonomy of the key United Nations organs in which these functions are undertaken. The choice of organizations selected as examples was based on their core mandates and the scope of their activities. Efforts have been made to ensure that the major categories of United Nations organs have been covered and therefore the results of the review are representative of the functioning of the United Nations system.

The organs of the United Nations perform a wide range of functions that require varying degrees of scientific and technical input. These functions range from rule making to operational activities that involve project implementation at the local level. Some of the functions that are related to sustainable development include norm setting, guidance, and advocacy; research and development; assessment, monitoring, and reporting; operations, technical assistance, and technology transfer; and science and technology advice (which is either addressed through independent institutional arrangements or as part of the other functions).

In addition to focusing on United Nations organs that have primary responsibilities in the fields of water, energy, fisheries, and oceans, the Committee surveyed more than 40 United Nations organs and examined their institutional arrangements for providing science advice. Committee members also examined the records of the *Earth Negotiations Bulletin*, the most comprehensive records of conference reports on sustainable development available to the public. This information was complemented with published sources, although these were mainly general in nature and focused on the role of science advice in environmental negotiations broadly or the implementation of specific environmental regimes represented by international treaties. The published material provided insights into the functioning of the regimes but provided little information on the inner working of the organizations surveyed.

The discussion of Chapter 2 distills the experience of a number of international and national science advice institutions with membership from developed and developing countries. Via e-mail, the Committee asked a number of organizations what procedures they had adopted for developing science advice. These were the Inter-Academy Council (IAC), the Council of Academies of Engineering and Technological Sciences (CAETS), the Third World Academy of Sciences (TWAS), the European Council of Applied Sciences and Engineering (Euro-CASE), and the UK Royal Academy of Engineering (UKRAE). In addition, the UKRAE referred the chairperson to the procedures adopted by the UK Chief Science Advisor, which may be found on line at www.dti.gov.uk/ost/ostbusiness/july_policy.htm. Although not all contain all the elements to be shown below, the procedures of these organizations are generally consistent with each other, and with those of the US Academies, as well as with the general principles of science advice (Golden, 1991). The specific procedures employed by a sampling of these, as well as the UN Intergovernmental Panel on Climate Change, are described in Appendix II.

This information, together with the procedures and practices of the US Academies and the experience of the Committee members, formed the basis of the discussion of a model process for developing science advice, together with comments upon the various issues and problems. It is intended to be over inclusive, and it is not the Committee's view or recommendation that all the processes and procedures described must be followed in detail in all cases. Each organization will have to ponder the issues involved and adopt and adapt these prescriptions to their own needs. The material in Chapter 2 should, however, provide a guide to most of the problems and pitfalls that have been encountered by others, and some of the means by which organizations have solved the problems they present.

The term "science" is here taken very broadly to include areas such as health, agriculture, the social sciences, technology, and engineering. The Committee also recognizes that traditional knowledge systems are increasingly being recognized in fields such as biodiversity and resource conservation, and they play an important role as input into decision making on issues related to sustainable development (Posey, 1999). These systems can bring otherwise unknown empirical data to scientific attention, and practitioners can sometimes expedite the application of science and technology to local situations.

The problems of provision of science inputs to policy begin with understanding which questions are fit subjects for science advice, what form science advice should take, and how to distinguish balanced, authoritative, evidence-based advice from advocacy designed to advance specific viewpoints or interests. Because science advice involves time and resources, it is also necessary to know whether science advice is really needed, how it will be used, and how to manage a scientific advisory process.

Chapter 1 of this report outlines the development of science advisory systems internationally, and in the UN system in particular. It provides a starting point for a more detailed examination of functional and structural typologies of UN organs with a sampling of the existing science advisory mechanisms.

Chapter 2 explores ways in which organizations can establish and manage processes for generating and using science advice and reviews the experiences of science advice in international, regional, and national organizations. Important previous reports on science advice and the science advisory procedures used by a dozen academies of science and engineering were studied. From these, Chapter 2 extracts a template and global standard for analysis of science advisory mechanisms in the UN system.

Chapter 3 describes a functional typology sample of specific functions of the United Nations and how they are applied to the issue of sustainable development. The role of science advice where it exists is explicitly highlighted.

Chapter 4 illustrates the use of science advisory mechanisms in the UN system. It provides a structural overview of UN organs with a sampling of the science advisory mechanisms that exist in each type.

Chapter 5 offers recommendations for strengthening the institutional arrangements for science advice in the UN system, some of which could be implemented in the follow-up to the WSSD. These are offered in the belief that science and technology advice can make ever more effective contributions to the governance of the United Nations organizations and that there should be increased international attention to facilitating the access to and use of such advice.

This report has been reviewed in draft form by individuals chosen for their diverse perspectives and technical expertise, in accordance with procedures approved by the NRC's Report Review Committee. The purpose of this independent review is to provide candid and critical comments that will assist the institution in making its published report as sound as possible and to ensure that the report meets institutional standards for objectivity, evidence, and responsiveness to the study charge. The review comments and draft manuscript remain confidential to protect the integrity of the deliberative process. We wish to thank the following individuals for their review of this report: Peter Collins, Royal Society; Elizabeth Dowdeswell, University of Toronto; Gisbert Glaser, International Council for Science; Mohamed H.A. Hassan, Third World Academy of Sciences; Eduardo Krieger, Brazilian Academy of Sciences; Thomas Malone, North Carolina State University; Daniel Martin, Gordon and Betty Moore Foundation; Erling Norrby, The Royal Swedish Academy of Sciences; Yves Quere, French Academy of Sciences; Rustum Roy, Pennsylvania State University; Eugene Skolnikoff, Massachusetts Institute of Technology; Edith Brown Weiss, Georgetown University; and Keith Yamamoto, University of California, San Francisco.

Although the reviewers listed above have provided many constructive comments and suggestions, they were not asked to endorse the conclusions or recommendations, nor did they see the final draft of the report before its release. The review of this report was overseen by Alexander Flax, consultant. Appointed by the National Research Council, he was responsible for making certain that an independent examination of this report was carried out in accordance with institutional procedures and that all review comments were carefully considered. Responsibility for the final content of this report rests entirely with the authoring committee and the institution.

CONTENTS

Executive Summary   1

1 The Rise of Advisory Systems for Science and Technology   5
    Introduction, 5
    Science and international cooperation, 5
    Evolution of science advice
      in the United Nations System, 6
    Emerging trends, 10
    Conclusion, 11

2 Elements of Science Advice   13
    Introduction, 13
    Elements of science advice, 13
    Role of a science advisor, 13
    Knowing when science advice is needed, 14
    Stating the science advisory task, 15
    Identification and recruitment of a study committee, 15
    Balance of regions, disciplines, and views, 16
    Management of bias and conflict, 17
    Management of data; role of staff, 17
    The report, 18
    Consensus and dissent, 18
    Review, 18
    Delivery of advice, 19
    Implications of the process, 19
    Follow-up and impact, 20
    Conclusion, 20

3 UN Sustainable Development Activities and Their Science Advisory Processes   21
    Introduction, 21
    Science and sustainable development, 21
    Norm setting, 22
    Research and development, 24
    Monitoring, assessing, and reporting, 25
    Operations, technical assistance, and technology transfer, 26
    Conclusion, 29

4 Structure of Science Advice in the United Nations System Today   31
    Introduction, 31
    Office of the United Nations Secretary-General, 31
    Functional commissions, 33
    Programs, 35
    Conventions, 38
    Specialized agencies, 45
    Processes, conferences, and joint research activities, 49
    Allied activities, 53
    Non-state actors, 54
    Conclusion, 55

5 Findings and recommendations   57
    Introduction, 57
    Findings, 58
    Recommendations, 59

References   63

Acronyms   71

Boxes
Box 1.1: Science and technology diplomacy, 7
Box 1.2: Science and technology for sustainable development, 8
Box 1.3: History of science and technology advice in the United Nations, 9
Box 4.1: Science advice for marine resources conservation, 36
Box 4.2: Review process for IPCC reports, 40
Box 4.3: Science advice for fisheries management, 46
Box 4.4: Science advice on freshwater, 48
Box 4.5: Some major international activities on freshwater involving science, 52
Box 4.6: Science advice on marine pollution, 53

APPENDIXES

I   The United Nations System, 76

II   Procedures for the Preparation, Review, Acceptance, Adoption, Approval, and Publication of IPCC Reports, 77

III   *Modus Operandi* of the Subsidiary Body on Scientific, Technical, and Technological Advice of the Convention on Biological Diversity, 85

IV	Draft Resolution on the *Modus Operandi* of the Scientific and Technical Review Panel (STRP) of the Ramsar Convention on Wetlands, 90
V	Rules of Procedure of the InterAcademy Council, 97
VI	Procedures for the Production and Review of Proactive Academy Reports and Statements of the Royal Academy of Engineering, 99
VII	Study Procedures of the International Council of Academies of Engineering and Technological Sciences, 103

EXECUTIVE SUMMARY

Since the 1950s, international cooperation in science has grown. The United Nations has assumed a responsibility to advance human health, welfare, and development, while better managing and conserving the environment and natural resources. Because of the inherent technical aspects of this responsibility for sustainable development[1], the United Nations organizations have steadily increased their interest in scientific and technological issues. This has bred a need for science advice and reliance on experts from outside the organization. This report[2] addresses the ways in which the need for scientific input has been met, compares the methods to a composite set of good and tested approaches used for scientific input by other organizations, and makes some recommendations for directions in which the UN system might move.

The need for scientific advice in the UN system has been approached in different ways by different organs of the system and at different times. The United Nations structure includes a General Assembly, commissions, programs, research institutes, agencies, treaty bodies, forums, and conferences. Several of the organs of the United Nations are autonomous and respond to the countries through their governing bodies rather than to the General Assembly or other UN organs. The agency heads are responsible for alerting governments and governing bodies to emerging issues in their areas of jurisdiction, including scientific issues.

Science advisory mechanisms of one sort or another are found throughout the system. Many of the organs of the United Nations have set up advisory committees or processes to provide scientific input into decision making but there are no standard or generally applicable procedures that ensure quality and balance. A common approach is "conference diplomacy," relying on conferences, workshops, and expert group meetings to provide advice, with documents prepared by the secretariat or consultants; heavy weight is given to geographical representation in selection of experts, and there is no scientific peer review. Treaty organizations dealing with a variety of environmental and sustainable development issues have started to establish subsidiary bodies to bring science advice into their functions, but mainly in the context of negotiating positions. There also are a few quasi-independent science assessment processes in the United Nations system that provides advice to governments. These are surveys and analyses of

---

[1] Defined as "development that meets the needs of the present without compromising the ability of future generations to meet their own needs." (WCED, 1987). A more recent definition considers sustainable development to be "the reconciliation of society's development goals with its environmental limits over the long term," (NRC, 1999a).
[2] See the Preface for an explanation of the background and task statement for this report.

the status of one or more important global problems, often with recommendations for international cooperative action. One of the best known is the Intergovernmental Panel on Climate Change (IPCC), which is often held up as a model for such scientific assessment processes.

A general problem common to all the UN bodies is the inherent difficulty of achieving both scientific credibility and influence on the political process. Scientific credibility normally rests upon the expertise, independence, and objectivity of the body issuing an opinion. Bias and conflicts of interest are considered defects to be eliminated or balanced. The scientific community places great stock in peer review, where reports or recommendations of a group of scientists are reviewed and criticized, usually anonymously by other scientists of equal expertise and standing, and the original group is expected to respond to the their criticisms.

The political process rests on quite different foundations. The UN organs are representative bodies, and rely for decision making on the interplay of national interests. Expertise is frequently given less weight than balance of interests, the opposite of the independence prized by scientists. Weight is also given to geographical, economic, and even religious balance in advisory bodies. Since most scientists work in the Western industrialized countries and Japan, a recruitment of the world's best experts in many technical areas will result in an expert group with a majority of Westerners; such a body is given little credence within the UN system.

IPCC clearly stands out as a remarkable innovation in science advice in the United Nations system.[3] Being an intergovernmental process, it illustrates the strengths and difficulties of integration of the science advice with policy making. The fact that IPCC relies on peer-reviewed research for its assessments discriminates against scientific input from countries whose scientists do not have opportunities to release their studies in peer-reviewed journals. Similar imbalances exist in regard to the distribution of research activities in general. The processes that allow for interactions between scientists and policy makers help to reduce these concerns, but they raise other questions related to the objectivity of the scientific advice. The use of consultancy reports, workshops, and other tools of conference diplomacy result in a mingling of political interests and scientific assessments to levels that may undermine the credibility of the outputs.

The absence of clear procedures dealing with science advice generally exposes United Nations staff to political influence, which compromises the management of the advisory process. While many of the reports may be excellent, their basis in political processes often casts a shadow of doubt over their scientific standing. Some UN organizations are instituting measures to improve review processes and are developing rosters of experts. But there is still dependence on a limited number of consultants who often have considerable individual influence on the outlook of the organizations that they serve.

To a scientist, science advice has a very concrete meaning. There is usually a powerful consensus among working scientists regarding the current state of knowledge in any field, and there are established procedures that are used by the scientific community to determine and express that consensus. The scientific consensus evolves with time through

---

[3] See Appendix II for a description of IPCC procedures.

experiment, discovery, and new theoretical ideas, and what is considered good science advice mirrors this evolutionary process, taking account of change and continuing uncertainty. A balance must be found that retains scientific integrity and still provides policy influence.

Reflecting the nature of science itself, science advice involves certain principles and procedures. These generally involve the participation of balanced multidisciplinary expert panels selected to study specific problems, with heavy reliance on peer review of consensus reports and explicit exposure of areas of uncertainty, disagreement, and dissent. A partial solution is to have the science advisory process of each UN body managed by an appointed science advisor within the directorate who can provide interpretation for the political bodies of credible science advice and guidance for the scientists on the policy issues. This means that a science advisor is not simply one who "tells truth to power," in part because no science advisor can possess more than a small part of the knowledge that is relevant to most complex scientific questions. Instead, the role of the advisor should be to organize the process and serve as a link between decision makers and the scientific community. The science advisor can formulate the questions, recruit the experts, oversee the process, organize the peer review, and then advise the executive how to make use of the advice in the light of other legitimate inputs and interests. This will effectively create a new service within the secretariat to serve as a buffer and intermediary between the strict procedural demands of a credible science advice system and the give and take of the political process.

There are several institutions, national and international, that have developed science advice procedures based on independence and peer review. The committee has surveyed many of these institutions' procedures, and descriptions of their processes appear in the annexes. This information, together with the procedures and practices of the US National Academies and the experience of the committee members, formed the basis of the discussion of a model process for science advice presented in chapter 2 of this report. It is intended to be over inclusive; it is not the committee's view or recommendation that all the processes and procedures described must be followed in detail in all cases. Each organization will have to ponder the issues involved, and adapt these prescriptions to their own needs. The material in Chapter 2 should, however, provide a guide to most of the problems and pitfalls that have been encountered by others, and to some of the means by which organizations have found solutions.

In this report, some actions are proposed to enhance the availability, value, and use of science advice within the United Nations system.

Recommendation 1: Governing bodies of the United Nations that have substantial responsibilities for implementing sustainable development programs should each create an Office of the Science Advisor or equivalent facility or organizational function appropriate to its mandate.

The science advisory function should be within the office of the Secretary-General, Director-General, or Executive Secretary of the organ or conference and should serve the governing body of the organization through the Secretariat. These bodies include the governing bodies of specialized agencies, and the governing bodies of specially convened international meetings, such as the 1994 UN meetings on population in Cairo the World summit on Sustainable Development in Johannesburg, and the World summit on the Information Society to be held in 2003 and 2005 in Geneva and Tunis, respectively,

The purpose of this recommendation is to improve the current processes by inserting a person or organizational mechanism to assist in the recognition of policy issues that require science advice, in the formulation of the science questions in obtaining the science advice, and in interpreting the resulting scientific advice and in using it in the ensuing policy process.

Recommendation 2: Each such science advisory facility or organizational mechanism should adopt an appropriate set of general procedures based on those described in this report, adapted to any special circumstances of the organization. These procedures should be widely publicized within the corresponding diplomatic and scientific communities. The purpose of this recommendation is to ensure that the UN Organizations move toward reasonable uniformity of science advisory practices based upon the best practices of the word scientific community.

Recommendation 3. The United Nations should help member states to strengthen their own scientific advisory capabilities, and it should recruit scientists associated with these national capabilities for UN scientific advisory functions. The United Nations will be better able to use scientific advice when all nations have the capability to participate fully in its scientific advisory processes.

Recommendation 4. To complement their internal scientific advisory processes, chief executives and deliberative assemblies, separately or in cooperation, should commission science policy advice from established independent organizations that follow procedures similar to those described here.

Recommendation 5. Assemblies and other deliberative bodies should make greater use of scientific assessment mechanisms, such as the IPCC, that have the transparency and credibility of a scientific process. Scientific assessment mechanisms provide a good model to be considered for other nonscientific, deliberative, and advisory processes.

Chapter 1

The Rise of Advisory Systems for Science and Technology

Introduction

International cooperation on scientific and technological matters has a long history that predates the United Nations. When the UN technical assistance programs were created, many of the cooperation mechanisms were merged into specialized United Nations agencies such as the World Meteorological Organization (WMO) and the United Nations Educational, Scientific and Educational Organization (UNESCO). In addition, the United Nations is now emerging as a central forum for harmonizing international practices in a wide range of fields related to science and technology.

Science and international cooperation

International cooperation in the planning and execution of scientific observations and experiments was raised to a new scale in 1957-1958 with the International Geophysical Year (IGY), an observation and research program of 18 months duration. Scientists working in 67 nations, by prior international agreement, made meteorological, seismic, and upper atmospheric measurements—many of them simultaneous and coordinated—at observatories that extended from the North Pole to the South Pole. They explored the oceans and traversed the Antarctic continent. Data were freely shared and archived in three centers that made them accessible to scientists everywhere. The remarkable aspects of the IGY, in retrospect, were the broad scale of its vision and the fact that it was planned and took place at the height of the Cold War, proving that scientific cooperation could transcend political tensions and national boundaries.

The success of the IGY led to a great number of collaboratively planned research programs, initially in the geophysical sciences, that were soon extended to other areas that transcend national boundaries, such as space and ecology. Some international scientific organizations, including the Scientific Committee on Oceanographic Research (SCOR), the Scientific Committee on Antarctic Research (SCAR), the WMO, and the agencies carrying out cooperative planning in the hydrological sciences that formed the basis of the International Hydrological Decade, have carried the planning and coordination of cooperative research programs forward over several decades. Although the Committee on Space Research (COSPAR) was formed alongside SCOR and SCAR in the late 1950s, cooperation in planning in the area of space research has been less successful at times because of continuing tensions between the Cold War powers over the civilian and military, or dual, use of space technology. But, in recent years, the promise of true cooperation in space has also expanded as the Cold War opponents have become partners and more nations have attained the technical capability to build and launch satellites and share in both the excitement of discovery and the cost of space-based platforms.

As the world has gained increased experience in international planning of research programs, there has appeared a new form of cooperation explicitly related to policy questions, often described by the term "scientific assessments." These are surveys and analyses of the status of important global problems, often with recommendations for global cooperative action. Some of today's international science organizations— intergovernmental and nongovernmental—are the products of scientific assessments and date their origins to the programs started in the post-IGY period. Environment, ocean, and climate assessments that are updated periodically to provide information on trends or changes have been particularly useful to decision makers (IPCC, 2001).

Evolution of Science Advice in the United Nations System

Since the 1950s, the United Nations organizations have steadily increased their interest in scientific and technological issues and their need for science advice, especially for the planning of sustainable agricultural, forestry, and fisheries programs and for capacity building in developing nations. But the trend for use and acceptance of science advice from the 1960s to the present has not always been positive. Indeed, there simultaneously have been advances in the development of science advisory mechanisms and setbacks in the use and application of the resulting advice.

There was an early optimism in the 1970s that "technology transfer" was the key to the solution to underdevelopment and that it could be readily accomplished. It was thought that solutions to developing country problems could be easily taken "off the shelf" from industrial nations, and that the owners of the intellectual property would be happy (or could be forced) to share their knowledge with earnest and deserving colleagues in developing countries. Several UN organizations based on this premise were established in the 1960s, some of which were attached to the United Nations Economic and Social Council (ECOSOC), but they were not successful and disappeared or were reorganized and redirected (Sagasti, 1984; 1999).

The UN Conference on Science and Technology for Development (UNCSTD) in Vienna, 1979, marked a conceptual shift in the views of both industrialized and developing nations (Wilkowski, 1992). The meeting brought into the open many of the key issues, and it forced many in developed countries to confront seriously the valid aspirations of developing country scientists and governments. However, even serious consideration did not in most cases lead to agreement, and many imaginative UNCSTD creations, such as a financing system for science and technology for development, did not endure. UNCSTD sharpened the conviction in industrialized nations and developing nations alike that the building of endogenous scientific and technology capabilities in developing nations was central to their future prosperity. Growing recognition in the industrializing nations of the importance of market forces and the role of the private sector also heightened interest in the contributions of science and technology.

> **Box 1.1: Science and technology diplomacy**
>
> The role of science and technology in international diplomacy is increasingly being recognized as a key element in the functioning of the United Nations system. This trend is illustrated by the emergence of programs that specifically focus on the interface between science and diplomacy. For example, in July 2001, the United Nations Economic and Social Council adopted Resolution 2001/31, which established the Science and Technology Diplomacy Initiative under the leadership of the Secretary-General of the United Nations Conference on Trade and Development (UNCTAD), following a recommendation of the UNCSTD in consultation with the Secretary-General of UNCTAD.
>
> The resolution asked UNCTAD to "develop special programs and organize workshops...to contribute to ongoing programs for training scientists, diplomats and journalists in science and technology diplomacy, policy formulation and regulatory matters to assist developing countries, in particular least developed countries, in international negotiations and international norms and standard-setting."
>
> In another effort, the United Nations University, through its Institute for Advanced Study (UNU/IAS), has launched a research program on biodiplomacy as an effort to clarify the relationship between international diplomacy and the biosciences. Other decisions in the United Nations also illustrate the growing recognition of this topic. For example, United Nations Secretary-General Kofi Annan has appointed former Costa Rican President Jose Maria Figueres as his special envoy on information technology, a task that involves addressing diplomatic issues associated with the role of information technology in society.

Interestingly, one of the major arguments for creating viable scientific communities, even in countries that were unlikely to be major contributors to new scientific and technological knowledge, was the need of every government for science and technology advice. In other words, for developing countries it appeared that a main purpose of science was to produce scientists, multipurpose assets whose advice is needed. Every government has to make decisions regarding education, procurement of technology, agricultural development, health, industry, natural resources, and mining that require reliable knowledge of the state of the art in science and technology. But, even so, the UN and most of its subsidiary organizations did not evolve an effective science and technology advisory system at that time, either for themselves or for their member countries, although, as described in Chapter 2, some advisory mechanisms in the UN system have played useful roles. Over the decades, the use of "science advisers" has increased, and the countries themselves often select delegates to the assemblies who have considerable understanding of complex science and technical issues in environment, agriculture, and fisheries.

> **Box 1.2: Science and technology for sustainable development**
>
> The role of science and technology in sustainable development is clearly articulated in the reports of the 1992 United Nations Conference on Environment and Development in Rio de Janeiro. Principle 9 of the Rio Declaration on Environment and Development codified this theme by calling on States to "cooperate to strengthen endogenous capacity-building for sustainable development by improving scientific understanding through exchanges of scientific and technological knowledge, and by enhancing the development, adaptation, diffusion and transfer of technologies, including new and innovative technologies." Agenda 21, the work program of the conference, identifies science and technology (together with finance) as the central means for implementing the transition toward sustainable development. This recognition is reflected in nearly all the 40 chapters of the work program.

The 1963 United Nations Conference on the Application of Science and Technology for the Benefit of Less Developed Countries provided the early beginning of discussions on science and technology advice. The conference was convened on the assumption that technology transfer from the industrialized to the developing countries could spur rapid economic transformation in regions most in need of it (Juma, 2002a).

In the late 1960s and early 1970s, after the UN had established its original science and technology advice system, global development theories shifted from those of modernization, whereby the problems of the world's poorer nations were to be solved through technology transfer and the dissemination of technological advancements, to a pattern of economic and social restructuring. This shift was associated with the growing recognition of the role of science and technology for international competitiveness as well as development. Multilateral lending agencies, rather than scientific research institutions, became the centers of development initiatives. The task of development was defined largely in terms of macroeconomic theories embodied in structural adjustment programs (Binswanger, 2001). Developing countries pressed for greater access to technology, as well as investment in science and technology capacity development.

Box 1.3: History of science and technology advice in the United Nations

1945	United Nations officially chartered at conference in San Francisco, CA, USA
1963	United Nations Conference on the Application of Science and Technology for the Benefit of Less Developed Countries. This conference marked the origins of the UN focus on science and technology as useful and important tools for development. It was based on the view that the developing countries could "leapfrog" across generations by adopting technologies developed in the industrialized countries. The conference resulted in a multifaceted approach to science and technology advice: 1. The Committee on Science and Technology for Development (CSTD) was to be a new component of ECOSOC. 2. The Advisory Committee on the Application of Science and Technology for Development (ACAST) was made up of experts in the various fields of science and technology and was created to provide science advice to CSTD and other UN bodies. 3. The Office of Science and Technology (OST) was created as a part of the UN Secretariat to support both of these groups and assist in the implementation of the advice.
1979	United Nations Conference on Science and Technology for Development. This conference was convened in response to criticism of the science and technology advice system put in place in 1963. The resulting Vienna Program of Action called for an updated version of the existing mechanism and placed emphasis on the need for capacity building and technology transfer. 1. The Intergovernmental Committee on Science and Technology for Development (IGC) was to replace the CSTD for setting science and technology directives. 2. The UN Advisory Committee on Science and Technology for Development (ACSTD) replaced the ACAST. This new group was comprised not only of scientists, but experts from government and business sectors as well. 3. The OST was upgraded and renamed the Centre for Science and Technology for Development (CSTD). 4. A funding mechanism was put in place and named the UN Financing System for Science and Technology for Development (UNFSSTD).
1993	The IGC, the Centre for Science and Technology for Development, and the ACSTD were abolished and in their place the General Assembly created the Commission on Science and Technology for Development. This new commission was to be supported by the science and technology divisions of the UN Conference on Trade and Development. The Commission was relocated from New York to Geneva.
2002	The Commission is still relied upon for science and technology directives within ECOSOC and the UN at large. Other bodies of the UN have created science and technology advisory mechanisms after the ECOSOC model, making modifications as they suited the purpose of each specific UN body.

SOURCE: Juma (2002a), adapted from Sagasti, F. "Science and Technology in the United Nations System: An Overview". Report prepared for United Nations Development Programme, New York, 1999.

This shift was reflected in a changing agenda for the UN system. As industrial countries continued to press for macroeconomic reform, developing countries sought to put in place policies aimed at attracting foreign investment. The UN science and technology advisory apparatus, which had been designed to promote technology transfer to developing countries, became increasingly irrelevant to the dominant patterns of international relations. As the gap between science and technology experts and policy makers grew, critics of the UN system called for a stronger social component together with a broader, more interdisciplinary approach to science advisory processes (Sagasti, 1999). While development policy stressing structural reform and investment was being promoted through institutions such as the World Bank, the United Nations continued to champion the traditional role of science and technology for development, and much of the diplomatic activism shifted to issues like intellectual property protection. The growing diplomatic conflict in these areas led to the convening of the United Nations Conference on Science and Technology for Development in 1979 in Vienna (Wilkowski, 1982).

The Advisory Committee on Science and Technology for Development included experts from business, government, science, technology and other areas. The new science and technology advisory mechanisms had the capacity to deal with crosscutting issues of development and the importance of science and technology for understanding and mitigating or advancing these issues. The UNFSSTD, the funding mechanism that was put in place by the UN in 1979, was an important improvement over the old system. It was to be a voluntary fund aimed at capacity building in science and technology and enhancing technology transfer.

These developments were complemented by other important strands related to science advice on environment and natural resources, championed through institutions such as the United Nations Environment Programme (UNEP) and specialized agencies such as UNESCO and the Food and Agriculture Organization (FAO). These organs would later be the locus of a wide range of scientific activities that not only informed decision making but also helped to create a new generation of environmental treaties that incorporated some of the principles of science advice. UNEP played a particularly important role in bringing science to bear on decision making. For example, the conventions dealing with climate change, ozone depletion, biodiversity, desertification, and chemical pollution have clear provisions for scientific assessments and technology transfer similar to those advocated for the general development purposes. *The global environmental movement has been an important source of experience on the use of science advice in international diplomacy.*

Emerging trends

The recognition of the growing pervasiveness of science and technology and the need for wise and informed advice on scientific and technical matters by all governments led the world's academies of sciences to form two collaborative organizations during the past decade. The InterAcademy Panel (IAP) is an organization of approximately 80 national and regional science academies of the world. It was organized in 1993 to prepare common statements on major issues of international concern, and its central focus has become mutual support among member academies in building capabilities to provide science-based advice to national and international policy makers. The first example was a statement on population growth, prepared for the United Nations Population Conference in Cairo in 1994 (NRC, 1994).

In 2000, the IAP created a second organization, the InterAcademy Council (IAC), to undertake science policy studies and offer real science advice to governments and organizations. The IAC is intended to advise intergovernmental organizations on policy questions with a high scientific and technological content, bringing to bear the world's best experts drawn from all regions. UN Secretary-General Kofi Annan has welcomed the IAC with a request for an urgent study on how agricultural productivity in Africa can be increased.

Conclusion

Over the course of the twentieth century, science became a truly international activity through the creation of organizations such as the International Council for Science (ICSU) and its disciplinary unions. In the last half of the century, scientists mobilized themselves and their governments for the initiation of global-scale studies. More recently, the world's science academies have created mechanisms to provide science advice for governments and international organizations, and ICSU is also moving in that direction.

The United Nations system has made use of science advice since its founding. Viewed over the five decades since the creation of the UN, the incorporation of science advice into decision making has steadily increased, albeit with setbacks from time to time. *Today, science, engineering, medicine, and health services are central to the resolution of most of the social and political issues that confront the world community of nations. It is imperative, therefore, that the UN system continues to strengthen its science advisory mechanisms.* The next chapter will describe the nature of science advice and offer some principles derived from the best practice of the world's scientific organizations.

Chapter 2

Elements of Science Advice

Introduction

To evaluate the quality and effectiveness of the science advice prepared and utilized by the UN and its component organizations, it is necessary to understand the meaning of science advice and how leading scientific organizations practice it. This material was collected by means described in the Preface and is here generalized so that it can be adapted and adjusted to the needs of specific users. It may be kept in mind as an aid in evaluating the science advice procedures and practices of the UN system, as described in Chapters 3 and 4.

Elements of science advice

There is usually a powerful consensus among working scientists regarding the current state of knowledge in any field, and there are established procedures that have the confidence of the scientific community to determine and express that consensus. The consensus knowledge in science evolves with time through experiment, discovery, and new theoretical ideas, and what is considered good science advice mirrors this evolutionary process, taking account of change and uncertainty. Like science, science advice is not so much a body of information as a procedure, which utilizes the processes of the scientific culture to decide questions of science. These processes lean heavily on peer review and the establishment of consensus, with the explicit exposure of areas of uncertainty, disagreement, and dissent.

Role of a Science Advisor

A science advisor cannot be one individual who "tells truth to power," in part because no imaginable science advisor can possess more than a small part of the knowledge that is relevant to most complex scientific questions. The role of the advisor is to serve as a link between decision makers and the scientific community. The science advisor's key value is the ability to know how science works, and to be known and trusted in the scientific community to ensure that the process of science advice involves a broad perspective and produces the best balanced advice possible, with explicit explanation of its uncertainties and remaining unknowns.

A science advisor[1] to a senior decision maker cannot be expected to carry in his or her head all of the knowledge necessary to advise on every problem, or even all knowledge of the sources of knowledge. The rapid rate at which science and technology are evolving makes it more difficult to rely on the heroic expertise of one individual. This is particularly true in the United Nations system, given the diversity of countries, conditions, and scientific and technological priorities. What is required, rather, is a sound knowledge of the scientific process through practical, productive experience in a scientific subject and broad familiarity with the national and international organization of science. The advisor should be known among scientists and should be well regarded, if not necessarily eminent, although eminence can be a significant advantage.

It is a further advantage if the career of an advisor has involved contact with a variety of scientific disciplines, some experience in management and administration, and some contact with public policy. The ability to maneuver at the boundaries of science and general policy is important, as is the ability to deal with a variety of people and professions.

The real task of the science advisor is to serve as an intermediary to engage the broad scientific community in the service of the organization or the decision maker. The science advisor, or the office of the science advisor, must be able to set in motion the science advisory process described below. This may include assisting the policy maker(s) to see the relationship of policy problems to science issues, and assisting them to decide what science information and advice may be helpful. The science advisor must help to interpret the policy-making world to the scientific community and help interpret the science advice to make it most useful in the policy world.

It is important to distinguish the advisory group or study committee that is convened to provide advice on one single question from the more general science advisory board or standing committee that may advise policy makers on issues related to science and technology. In this chapter we will deal with the first kind of group, which we shall call a study committee for clarity and distinction.

Knowing when science advice is needed

Many policy problems that are apparently unrelated to science have scientific aspects, for which decision makers might profit from science advice. This may not always be clear unless someone familiar with the science is involved early in problem formulation. Much of the problem formulation in the United Nations is provided by government representatives and by the UN monitoring, assessment, and reporting systems. Several flagship reports of the United Nations, such as UNEP's *Global Environment Outlook*, UNDP's *Human Development Report* and UNESCO's *World Science Report,* highlight emerging issues whose solutions may require science advice. Science advice may also be required to deal with controversial issues upon which governments are planning common actions or are being requested to allocate resources. Other areas that benefit from science

---

[1] The discussion of the role of science advisor applies equally well to the chairman of a scientific advisory committee. It is not generally good practice for a decision maker to chair his own advisory committee. That is similar to a defendant acting as his own lawyer; he loses the option of rejecting the advice (without losing the confidence of the committee).

advice involve conflict that could threaten peace and security. For example, advice regarding the use of satellite imagery and global positioning systems is being used to settle border disputes between countries (NRC, 2002). The use of science advice early in the policy process may make the policy problem easier to attack.

It is the experience of the committee that narrowing the scope of a problem (smaller "intellectual bandwidth") often makes it more easily solved. *To the extent that science and science advice can help to narrow the scope of policy problems, or to divide them up into more manageable pieces, the probability of successful problem solution may be increased.*

Furthermore, resources often can be saved by using information available publicly or from private sources that can shorten the process and reduce the need for a formal science advice process. For example, the need for technical training or scientific education is universal, and the needs of one organization or country may be met by information already prepared by and for others. Not every problem requires a unique solution.

Stating the Science Advisory Task

The problems faced by UN policy makers are generally not inherently scientific problems. To be amenable to science advice, the problems must be stated in a form that identifies their scientific or technological elements. For example, a farmer must decide which crops to plant; in order for scientific experts to be of any help, the question must be expressed in terms of soil characteristics, climate, water, markets, and costs. The farmer receiving the advice will evaluate it in the context of his culture and economic situation and those of his neighbors before making a final decision. Similarly, a policy maker may want to know whether a certain organization regulating fisheries is performing effectively; for science advice, this must be translated into terms of target setting, data gathering, methodologies, monitoring, and results. The policy maker may then factor this assessment in with other legitimate political and economic considerations in making a budgetary decision.

Translating a policy question into one amenable to the advisory process requires consultation between those who need the advice and someone who understands how to do the initial formulation of the task so that it is clear on which issues the advice and advisors might help. It is important to identify at the beginning of the process the principal user or users of the advice, those who will be its primary recipients. These may be senior decision makers or their representatives—those who will take action based upon, or influenced by, the advice. The product of the consultation between the users and the science adviser is the statement of task, the instructions to the study committee containing the concrete questions to be answered.

Identification and recruitment of a study committee

The first step after the users have approved the statement of task is to recruit a study committee. This is an ad hoc group, generally convened solely to answer one question, and the members will be recruited precisely on the basis of their ability to contribute to the answer to that question. This step requires knowledge of the landscape of science and scientists and knowledge of possible sources of reliable and useful advice. The entire study might be contracted out to an independent organization like the InterAcademy Council or ICSU. If the advice is to be prepared in-house, outside experts will still play a key role. Persons selected may be employed at scholarly institutions, in industry, in private practice, or in national or international organizations. These must be selected on the basis of their expertise, as individual members, and should not "represent" their institutions, although their institutional knowledge may be valuable to the process. *In the United Nations context, this requirement would demand separation between expert advice and diplomatic representation.*

Issues of credibility frequently determine the eventual usefulness of the advice, and the selection process used will be important to the study committee's credibility. Policy makers and the public are frequently unable to assess the credentials of individual scientists, but they are likely to be influenced by the reputation for scientific integrity of the appointing institution and by the transparency of the process. In this key matter of credibility, there is special value in selection of members of advisory bodies by the scientific community itself, or by clearly scientific merit-based processes. Examples of institutions with such processes include various science academies and national and international science funding organizations. For example, the British Royal Academy of Engineering posts the proposed members of its science advisory committees on its public web site for comments by its members and others who are interested.

Balance of regions, disciplines, and views

In recruiting committee members it is important to include different kinds of experts who might usefully contribute to the advice. They may be representatives of different geographical regions, as often required in the United Nations, or scientists whose specialties are on the margins of the specific scientific question. The capability of scientists with differing views is usually evaluated on the basis of their previous work published in peer-reviewed journals.

It is often useful to include members in the study committee who have general competence in science but may not be experts in the specific areas to be discussed. These seemingly "inappropriate members" of the group, if well chosen, are able to ask questions and make useful suggestions that are unlikely to emerge from a specialized discussion. *It is possible to be inclusive of a range of relevant and responsible scientific opinion without including opinion that does not have any basis in data, observation, or logically developed theory. There may be legitimate reasons for the policy makers to take such views into account, but they should not be confounded with science advice.*

One of the key issues facing science advice in the United Nations is the uneven distribution of scientific and technical capacity among countries and the limited resources available to developing countries to support the participation of their nationals in international science advisory activities. Strengthening science advice in the United Nations will require simultaneous strengthening of science, and even science advisory institutions, in developing countries and countries with economies in transition.

Management of bias and conflict

It is frequently difficult to find study committee members who are sufficiently knowledgeable about the subjects involved and are at the same time unbiased (i.e., have no previously published or publicly expressed opinions or conclusions on the subject) and have no conflicts of interest (i.e., have no likelihood of personal gain or loss depending on the outcome).[2] One option is to use necessary experts who have conflicts of interest, not as members of the study committee itself, but as consultants to the committee, or invite them to speak to the committee in open meetings. This may allow the committee to receive the benefit of their knowledge without the appearance of bias in the advice itself. In other circumstances, a large fraction of the expert population may have bias or personal interest, and it may be necessary to include them as committee members, but in a balanced way. In the burgeoning field of genetic engineering, it is often found that nearly all experts competent to serve have already declared their views or participation in profit-making entities. In such cases, the specific biases and conflicts should be known to all who participate, to those who have a stake in the results, and to those who receive the advice. Where biases and conflicts are unavoidable, a balanced group and a policy of transparency is the best approach to getting credible, objective advice.

In any case, the deliberations of a study committee always should be shielded from external political influence. The members should be operating as scientists in their personal capacity, and not subject to external instructions, threats, or rewards that might depend on the conclusions they draw.

Management of data; role of staff

Most scientific questions cannot be answered entirely with the knowledge and data in the possession of the committee members alone, however well they may be chosen. The consideration of externally supplied data, theory, analyses, interpretations, and opinions is necessary. It is generally advisable to obtain this "outside" material in an open and public way, for example, by publishing lists of consultants and speakers heard by the committee and of references cited.

Management of data and information requires a well-trained staff to serve the study committee. Staff can carry out extensive review of the literature, commission analytical papers on particular aspects of the topic, arrange expert testimony to the committee, and convene workshops where a variety of experts may present data and express their views.

In addition to the competence of the staff, care should also be given to its independence. A large number of United Nations employees are either seconded by governments or

---

[2] In diplomatic forums, governments often bring to negotiations scientific expertise that supports their negotiating positions. In this respect, bias is inherently built into the process. Sovereign interest is naturally a driving force in international diplomacy, but science advice within the UN system must be kept separate and distinct from the diplomatic arena.

maintain close contacts with their home countries, and candidates for senior positions in the United Nations require the endorsement of their home countries. This political environment could introduce bias or perception of bias in staff performance. This has in some cases led to government distrust of documents prepared by secretariat staff, and it is not uncommon under such circumstances for government delegates to generate their own documents. This is particularly common in treaty negotiations where decisions are binding on governments. The principles of the management of bias, and of establishing clear and transparent procedures, must therefore also be extended to staff.

Generally, it is desirable to have the entire information gathering process as open as possible. Openness allows those interested to know the inputs made by others, and permits all interested parties to provide input, in person or in writing. However, there must always be opportunity for the committee to meet for discussion and debate without the presence of outside observers, sponsors, or interested parties. A distinction must be made between the information–gathering process, which is open to outside input, and the deliberative process that must be kept closed and free of outside influence. Aside from the technical and methodological issues involved, the issue of public confidence in the results must be paramount.

The report

The product of the science advice process should be in the form of a report. *A formal report is the mechanism by which it can be demonstrated that all members of the committee had an opportunity to present their views, that the recommendations are based on data and the best scientific knowledge at the time, and that outside views were taken into account.* A member or members of the committee, each writing sections of particular interest to them, the chair, or staff may draft the report. In any case, the study committee as a whole must approve the resulting draft.

Consensus and dissent

It may happen that, after reasonable efforts to produce conclusions and recommendations that achieve a consensus of the committee, one or more members are unable to agree to the draft. Sometimes the problem may be solved by including the details of the issue and the reasons for the dissent in the draft itself, thus achieving a balance to which all may agree. If this is not possible, signed written dissents may be attached to the draft. *If advice is to be useful it must be clear, and clarity may require statements of uncertainty and incompleteness where that is the state of knowledge at the time. Dissent may be part of providing this clarity.*

Review

The draft report should be reviewed by outside reviewers (scientific journals call them "referees" and research managers call them "peer reviewers"), some of whom are chosen for their expertise in the subject, and some because they have a good general view of the relevant technical subjects and are experienced in the science advice process. The purpose of review is not to "certify" the result but, rather, *to look for errors, lacunae, lack of clarity, missing information, etc.* The reviewers represent the universe of potential critical readers, and the review serves as "quality control" for the advisory process. It serves and protects the institution carrying out the advisory process.

Recruitment of reviewers is carried out in much the same manner as recruitment of the study committee, and bias and conflict of interest must be taken into account for the same reasons. However, the role of reviewers is such that a wider spectrum of views and biases may be appropriate for the reviewers than for the committee itself. Reviewers' comments often reflect national and other biases, and efforts should be made to ensure that the review process is not a substitute for introducing political negotiations into advice documents.

The committee should consider the comments of the reviewers and make appropriate changes in the documentary advice in response to them. Some institutions require that all comments be answered, whether changes are made to the work product or not. To avoid endless debates with reviewers, this can require some individual who has not been part of the process to be responsible for vouching for the satisfactory nature of the responses.

Delivery of advice

The final report is delivered to the principal users designated at the beginning of the process. The delivery is frequently accompanied by personal briefings to the users by the chair or other members of the committee. It is helpful if the chair or the designated members can continue to be available from time to time to answer questions and for further discussions and interactions with the users.

It is generally appropriate to publish or otherwise disseminate the report to legislative bodies, executive bodies, and the public. This will have clear benefits in credibility with parties interested in the matter.

Implications of the process

Generally, the most serious problems arise at the "front end" and at the "back end" of the process. The formulation of the task must be clear, relevant, and doable. The science advice must be delivered in an understandable form, with its implications for action (or inaction) clear. Those having the best and most detailed scientific knowledge are not necessarily best able to perceive its implications for policy, decision making, and action.

For this reason, it is useful for senior decision makers to have their own science advisor who has their trust and the resources to carry out the scientific advisory process as described. Because of the multidisciplinary nature of the process, a single isolated advisor without the resources to call upon experts in a variety of disciplines cannot provide science advice that will be accepted by the scientific community, whatever his or her credentials, or serve well the decision maker or the institution.

The participation of the science advisor in bringing science into the complexities of policy problem definition and into the interpretation of the science advice as it may impact upon policy can be very helpful in the implementation of the recommendations.

This can make the difference between successful use of science and the failure of an action because of scientific constraints. It is important that science be part of the senior policy process, where understanding its implications may be crucial.

Follow-up and Impact

Of course, even very good science advice is often not translated into policy, for a very wide range of reasons. As noted above, the scientific dimension, although very important, must compete with many other considerations. However, there are an impressive number of important cases in which solid scientific inputs have made critical contributions to policy. Well-known examples are scientific understanding of ozone depletion and an early study by the US Institute of Medicine which first called attention to the urgent need for awareness and prevention programs related to AIDS. The case of the IPCC is also noted elsewhere in this report. *The potential for high impact underscores the importance of publication, dissemination, and discussion of science advisory products with policy makers, groups that are affected by the policies in question, and the public.*

Conclusion

The elements outlined above are synthesized from efforts around the world to find ways of ensuring scientific credibility on the one hand and interactions between science and policy on the other. They may therefore provide a foundation against which to assess the role of science advice in the United Nations system.

Chapter 3

UN Sustainable Development Activities and Their Science Advisory Processes

Introduction

This chapter presents a review of the functioning of the United Nations system with emphasis on the role of science in sustainable development activities. As indicated in the Preface, the survey seeks to illustrate the role of science advice in informing policy on water, energy, fisheries, and oceans, which will require first an examination of the overall UN system. This chapter outlines the major functions and structure of the United Nations, and will try to demonstrate both the complexity of the United Nations system and the diversity of activities carried out by the organization. This complexity and diversity make it difficult to design and apply a single, uniform science advice system.

Many international organizations involved in monitoring, assessment, and reporting base their activities on a broad internal base of expertise of their staff and may not realize the importance of science advice. Some of the organizations such as IUCN and the World Bank have chief scientists responsible for providing science advice to the organization. The offices of the chief scientists provide a starting point for strengthening science advice through the adoption of appropriate procedures where they do not exist. Various organizations are presently reforming their science advice systems to incorporate variations of the elements outlined in Chapter 2, with the goal of seeking a balance between scientific credibility on the one hand, and effective interactions between science and policy on the other.

Science and sustainable development

The United Nations system plays a leading role in promoting the application of science and technology to sustainable development. Agenda 21 explicitly cites science and technology as key to the implementation of sustainable development goals, along with finances, human resources, and capacity building. Chapter 18 of Agenda 21 on freshwater resources, for example, states: "The development of interactive databases, forecasting methods, and economic planning models appropriate to the task of managing water resources in an efficient and sustainable manner will require the application of new techniques, such as geographical information systems and expert systems to gather, assimilate, analyze and display multisectoral information and to optimize decision making. In addition, the development of new and alternative sources of water supply and low-cost water technologies will require innovative applied research. This will involve the transfer, adaptation and diffusion of new techniques and technology among developing countries, as well as the development of endogenous capacity, for the purpose of being able to deal with the added dimension of integrating engineering, economic, environmental and social aspects of water resources management and predicting the effects in terms of human impact."

Similarly, Chapter 17 of Agenda 21 for oceans calls upon states to "cooperate in the development of necessary coastal systematic observation, research, and information management systems. They should provide access to and transfer environmentally safe technologies and methodologies for sustainable development of coastal and marine areas to developing countries. They should also develop technologies and endogenous scientific and technological capacities." The pervasiveness of the role of science and technology in Agenda 21 is accentuated in a call in Chapter 31 of Agenda 21 on governments to strengthen "science and technology advice to the highest levels of the United Nations, and other international institutions, in order to ensure the inclusion of science and technology know-how in sustainable development policies and strategies." This call is consistent with efforts to provide science and technology advice to the highest levels of governments (Golden, 1991).

This background provides the basis against which to relate the functions of the UN system to the role of science advice in sustainable development. The rest of this chapter describes a selected number of functions (norm setting; research and development; monitoring, assessing, and reporting; and operations, technical assistance, and technology transfer) and explores the extent to which scientific and technical information and advice are relevant to their effective execution.

Norm setting

A number of United Nations agencies are engaged in generating prescriptive statements or norms (guidelines, principles, standards, and rules) (Chayes and Chayes, 1995). This is one of the most important functions of the United Nations system. The norms are aimed at influencing the behavior of states, although the ultimate target is often to influence the behavior of institutions or individuals (Braithway and Drahos, 2000). The norms generated by the United Nations vary considerably in specificity. For example, the United Nations General Assembly and conferences focusing on specific themes usually generate guidelines that are general in nature and nonbinding in character, whereas some specialized agencies, such as the International Civil Aviation Organization (ICAO), the World Meteorological Organization (WMO), the World Intellectual Property Organization (WIPO), the International Telecommunications Union (ITU), and the International Atomic Energy Agency (IAEA), produce technical guidelines and standards that are more specific than the decisions of the United Nations General Assembly.[1]

A number of international conventions set specific rules on issues related to sustainable development. For example, the Convention on International Trade in Endangered Species of Wild Fauna and Flora (CITES) has specific rules and procedures for regulating international trade and is supported through scientific and technical input provided by the

---

[1] The International Organization for Standardization (ISO) is a non-UN agency that is devoted to standards setting in a wide range of activities that have direct implications for sustainable development activities in the fields of water, energy, fisheries, and oceans.

World Conservation Union (IUCN), the UNEP World Conservation Monitoring Centre (WCMC), and the World Wildlife Fund (WWF). CITES also relies on scientific and technical input provided by nongovernmental organizations (NGOs), whose participation has been central to the functioning of the convention (Curlie and Andresen, 2002; Keck and Sikkink, 1999). The World Trade Organization (WTO) is another example of an international body that is devoted to setting rules. As the level of specificity increases (i.e., from guidelines to rules), so does the demand for more specific scientific and technical information as well as institutional arrangements that ensure continuous review and harmonization of practices.

Whereas general state-of-the-art assessments are needed to support decision making in bodies such as general assemblies, the setting of standards and rules requires more detailed technical information that is provided through specialized scientific and technical committees or working groups. For example, the ITU produces over 200 new or revised standards a year through a series of technical working groups. This level of intensity in standards setting cannot be achieved through general conferences. However, efforts to establish major trends such as global warming require synthesizing available knowledge and establishing scientific consensus, as has been done by the Intergovernmental Panel on Climate Change (IPCC).

The effective functioning of rule-based organizations such as the WTO depends largely on the existence of institutional arrangements that make use of the latest available scientific and technical knowledge. The emphasis that WTO places on science-based decision making, especially on issues such as sanitary and phytosanitary standards in international trade, illustrates this point (NRC, 2000).

Debates over how to deal with scientific uncertainty in international trade continue to feature prominently in diplomacy. For example, the United Nations has sought to resolve some of the disputes over the safety of genetically modified (GM) foods by adopting the Cartagena Protocol on Biosafety under the Convention on Biological Diversity. The protocol is offered as an alternative for integrating environmental considerations into international trade practices through the promotion of the precautionary principle[2] (Gaugitsh, 2002). While advocates of this approach claim it would stimulate further research, others contend that the precautionary approach is not a science-based approach to decision making and suggest that its implementation would interfere with international trade (Marchant, 2002). It is notable that, although governments called for the use of the best available scientific and technical information to guide the biosafety negotiations, no systematic efforts were made to take stock of the available knowledge on the subject, and

---

[2] Article 10(6) of the Cartagena Protocol on Biosafety to the Convention on Biological Diversity articulates the precautionary principle as follows: "Lack of scientific certainty due to insufficient relevant scientific information and knowledge regarding the extent of the potential adverse effects of a living modified organism on the conservation and sustainable use of biological diversity in the Party of import, taking also into account risks to human health, shall not prevent that Party from taking a decision, as appropriate, with regard to the import of the living modified organism in question...in order to avoid or minimize such potential adverse effects." This is a contentious formulation because it does not define the scope of "taking a decision" and is therefore deemed to open the door for arbitrary action that does not demand the use of scientific evidence.

as a result much of the scientific input was provided through the direct contributions of government delegates (Gaugitsh, 2002). Divergent interpretations over the significance of existing studies on the safety of GM crops for human health, the environment, and socioeconomic systems continue to be a major issue of public concern (Gupta, 2000). The persistence of varied interpretations of the available information illustrates the need for scientific assessment to guide discussions and negotiations on major issues of international interest (Susskind, 1994).

Norm setting is one of the most important functions of the United Nations. This processes depends on continuous availability of scientific and technical information. While establishing general guidelines and principles can be achieved largely through the use of periodic scientific assessments, more technical efforts to change behavior through the use of standards and rules require continuous access to the latest available information. Whereas scientific assessments may require wide intergovernmental participation over longer periods, smaller technical working groups or committees often carry them out. *The growing complexity of economic activity and demand for broadening the scope of international governance require greater use of scientific information. This feature of global governance is likely to become more explicit in the future, leading to more frequent examinations of the way international organizations use science advice in their activities.*

Research and development

The United Nations carries out a wide variety of research activities, ranging from basic science to policy analysis. The specialized agencies of the United Nations, such as the International Atomic Energy Agency (IAEA), the World Health Organization (WHO), the Food and Agriculture Organization (FAO), and the United Nations Industrial Development Organization (UNIDO), are engaged in a variety of scientific research activities in their areas of expertise. Much of this work is carried out through partnerships with other research institutions around the world. The International Centre for Genetic Engineering and Biotechnology (ICGEB) located in Trieste, Italy and New Delhi, conducts research, provides services to member states and carries out training programs. More than 300 ICGEB researchers from 30 countries pursue inquiries into development problems such as malaria and hepatitis vaccines, study of human pathogenic viruses and human genetic diseases, and the genetic manipulation of plants (Juma, 2002a).

Another UN initiative, the Programme for Biotechnology in Latin America and the Caribbean of the United Nations University (UNU/BIOLAC), was established in July 1988 in Caracas, Venezuela. UNU/BIOLAC promotes biotechnology development in the Latin American and Caribbean region. Areas of focus include molecular biology, molecular pathology, genomics, industrial biotechnology, environmental biotechnology, and agricultural biotechnology (Juma, 2002a).

Equally important are social science research activities carried out in institutions such as the United Nations University, the United Nations Research Institute for Social Development, and the United Nations Institute for Training and Research (UNITAR). Much of the research carried out by these institutions is intended to provide policy recommendations to governments and the general public, and much of it touches on issues related to science advice. For example, the United Nations Conference on Trade and Development (UNCTAD) over the years has been a leading supporter of policy research related to the role of technology in development. The United Nations University, through its research centers, has also been a major player in policy research.

The United Nations is also involved in R&D activities through its support of the Consultative Group on International Agricultural Research (CGIAR), created in 1971. The aim of the CGIAR is to contribute to food security and eradicate poverty in developing countries through the use of research, partnerships, capacity building, and policy support. The CGIAR, as an association of public and private members, supports a network of 16 centers operating in more than 100 countries to mobilize science and technology to address hunger and poverty, improve human nutrition and health, and protect the environment. It operates at a budget of $320 million a year contributed by a consortium of donors. More than 8,500 CGIAR scientists and scientific staff are engaged in research to improve tropical agriculture. The CGIAR holds one of the world's largest *ex situ* collections of plant genetic resources in trust for the world community. It contains more than 500,000 accessions of over 3,000 crop, forage, and agroforestry species. The collection includes farmers' varieties and improved varieties and the wild species from which those varieties were derived. The collections have been placed under FAO administration.

Science advice is an integral part of the CGIAR, whose secretariat has a science advisor. As part of its reform program, the CGIAR has replaced its Technical Advisory Committee (TAC) with a Science Council (SC). The aim of the SC is to serve as the guardian of relevance and quality of science in the CGIAR and to advise the organization on strategic scientific issues relevant to its mission. The Council also functions as a strategic adviser to the Executive Council and its various committees. The SC focuses on ensuring that research throughout the system is peer reviewed. It consists of eight scientists with expertise in the biological, physical, and social sciences. The range of skills, as well as its size, will be regularly reviewed by the Executive Committee and adjusted to the needs of the organization.

Research and development has been an integral part of the international system since the creation of the United Nations, although its significance and size have changed over the years. The CGIAR, for example, played an important role developing new crop varieties to meet the food needs of the developing world. It is role, however, has been under constant review as its funding has declined. Despite these challenges, the CGIAR and other international institutions continue to undertake research of relevance to developing countries. *In order to support decision making, these institutions are seeking to enhance their science advice mechanisms. They are particularly seeking to strengthen their scientific credibility by strengthening peer-review mechanisms and widening the base for scientific input.*

Monitoring, assessing, and reporting

Various agencies of the United Nations have long tracked the environmental impacts of human activities, as part of their general role of monitoring trends, undertaking assessments, and reporting on progress. For example, monitoring of radio nucleotides arising from atmospheric testing of nuclear tests started in 1955 under the auspices of the United Nations Scientific Committee on the Effects of Atomic Radiation (UNCSEAR).

This work was based on data collected from a network of stations positioned around the world. It marked one of the earliest United Nations efforts to monitor anthropogenic environmental impacts. In 1963 the WMO launched the World Weather Watch, which formed one of the earliest networks for monitoring, processing, and reporting weather-related information (Gosovic, 1992). Since then, WMO has emerged as a key backbone of the global environmental monitoring system and has provided much of the basic data that have informed major international environmental policies (Davies, 1990).

Other institutions also monitor technological development, especially for purposes of setting performance and safety standards. The International Organization for Standardization (ISO), ICAO, and the ITU operate largely through national and regional institutional networks to set standards and make rules. WIPO, whose functions are limited to intellectual property information, also monitors and articulates trends in technology.

In the field of environmental management, the function of monitoring technological development is currently restricted to a few institutions working on specific technical problems such as developing substitutes for ozone-depleting substances. The ozone regime has a strong scientific and technological basis and includes a mechanism for providing financial assistance to countries to phase out ozone-depleting substances.

On the whole, the United Nations system has developed elaborate mechanisms for monitoring, assessing, and reporting on environmental trends. The system relies on a complex network of international, regional, and national institutions that collect environmental data. It also relies heavily on networks or epistemic communities that focus on specific areas of research (Haas, 1992). Institutions such as the International Council for Science (ICSU) have played an important role in supporting these networks. But with the emergence of the Internet and other communications technologies, a wide range of epistemic communities have emerged and are functioning without the support of coordinating organizations. The monitoring of technological developments, however, has not received as much attention in the United Nations system as have advances in environmental research.

The United Nations has an elaborate system for monitoring, assessing, and reporting environmental trends. Much of the information generated through these activities has had a significant impact on the emergence of the international environmental regimes. Not only has the information been used in setting new agendas and promoting international consensus, but also these systems are key to promoting compliance with environmental commitments. These activities are therefore important sources of information that feed into the various international science advice activities within the UN system.

Operations, technical assistance, and technology transfer

From project implementation to technical and financial assistance, international agencies take operational responsibility for social and economic issues such as development administration, children, refugees, food aid, environment, and population. UN technical assistance to developing countries over the years has been integral in these areas, particularly as coordinated at the country level by the United Nations Development Programme (UNDP).

An administrator, four deputy administrators, and a 36-member Executive Board that meets twice a year in regular session and once in special session oversee the largest program of the United Nations, UNDP. Its New York secretariat, along with offices in more than 130 countries, serves as its operational managers. Since reorganization of UNDP in 2000, a new financial instrument, the Thematic Trust Funds, has begun to finance UNDP projects. These funds enable donors to provide additional contributions for continuing work in specific UNDP practice areas. For the period 2001-2003, the energy trust has resources of $60 million, $51 million of which is allocated to country offices and $9 million to global and regional programs.

The energy trust of UNDP is managed by the Bureau for Development Policy (BDP), whose Sustainable Energy and Environment Division (SEED) carries out many of UNDP's energy program activities. The Capacity 21 program implements the goals of Agenda 21. In addition to utilizing UNDP Sustainable Development Advisors, country staff, and government counterparts, Capacity 21 maintains a network of about 12 global and regional advisors and 15 bilateral donors. Capacity 21 works closely with the Commission on Sustainable Development (CSD), whose energy and transport office is made up of 14 energy experts and a network of 200 consultants. UNDP is the second largest implementer, after the World Bank, of the Global Environment Facility (GEF) climate change projects.

The Energy and Atmosphere Program (EAP) supports a Sustainable Energy Knowledge Network, an e-mail network open only to UNDP headquarters and country offices. The EAP, with the World Energy Council (WEC) and the United Nations Department of Economic and Social Affairs (DESA), produced the *World Energy Assessment* for the 2001 meeting of the Commission on Sustainable Development (CSD-9). The report is also an informal input into WSSD in 2002. The *World Energy Assessment* evaluates the social, economic, environmental, and security issues linked to energy for every country and assesses the compatibility of different energy options with global objectives in these areas.

In addition to its operational programs, UNDP houses the Human Development Report Office (HDRO), which produces an annual report on a variety of themes related to human development. The *Human Development Report* is one of the most respected documents produced by the United Nations. It produces a ranking of the social and economic performance of countries based on a "Human Development Index." Recently, the report has focused on science and technology issues. In 2001, the Human Development Report was devoted to "Making Technology Work for Human Development." It outlined a wide range of policy options and helped to highlight the importance of science and technology for development. The staff of the HDRO, aided by consultants, prepares the report. The draft report is reviewed by a team of experts and launched with extensive media coverage. The report is not reviewed by governments and is the exclusive responsibility of the HDRO.

The United Nations has also played a key role in technology cooperation among countries. It has over the decades paid special attention to the need to transfer technology from the developed to the developing countries, especially for meeting human needs. This function is carried out through sectoral agencies dealing with issues such as health, agriculture, and industrialization. In addition, the United Nations promotes the transfer of technology among developing countries under the label of South-South cooperation. This work is coordinated through the UNDP.

One of the most successful United Nations efforts in technology transfer is the promotion of technologies that reduce the release of ozone-depleting substances (Tolba, 1998; Benedick, 1991; Litfin, 1994). This work is carried out though a complex network of institutions associated with the Vienna Convention on the Protection of the Ozone Layer, the Montreal Protocol on Substances that Deplete the Ozone Layer, and the Multilateral Fund set up to support technology transfer. The ozone regime has become a source of inspiration for the design of technology transfer programs. Its work is guided by the Technical and Economic Panel (Parson, Forthcoming; UNEP, 1999). One of the main reasons for the effectiveness of the UN in the ozone regime is the flexible institutional arrangement that promotes interactions between policy and science while maintaining the credibility of the latter. The ozone regime relied on leading experts and extensive peer-review procedures. In addition, the technical panels were broad in geographical coverage as well as in the composition of disciplines. The panels also relied on input from a wide range of constituencies, including government, industry, academia, and civil society. The panels cultivated an atmosphere of trust that enabled decisions to be guided by scientific consensus. Much of this was possible because of the existence of institutional structures that allow for scientific consensus building.

Another example of UN-based technology transfer activities is the work of the International Centre for Science and High Technology (ICS) set up in 1988 as an autonomous organ of the United Nations Industrial Development Organization (UNIDO). The aim of ICS is to promote sustainable industrial development through transfer of know-how and technology to developing and transition-economy countries. It carries out its activities through training courses, workshops, seminars, expert meetings, and project design. ICS works closely with small and medium-sized enterprises. It focuses on advancing the industrial competitiveness and investment climate by promoting technological innovation, building capacity, and promoting international cooperation. UNIDO has also set numerous other research and development programs in partnership with national governments.

Programs of the United Nations system are engaged in a wide range of operations that involve decisions on allocation of resources. Only a small proportion of these decisions are supported by systematic science advice. Although much of the technical knowledge needed for decision making is provided as part of the design and implementation of the various projects, there are no systematic institutional mechanisms and procedures within the organization to support and the advice provided. This is particularly critical for organizations whose mandate involves dealing with development issues that require considerable input from the scientific and technological community. *There are notable efforts to improve the functioning science advisory institutions where they exist. The improvements focus on enhancing scientific credibility (through measures such as peer review and inclusion of the social sciences) and introducing institutional improvements to support greater interaction between science and policy.*

Conclusion

The reforms identified above show the growing efforts of United Nations organs to balance scientific credibility with policy involvement. These findings are consistent with evidence from the literature (on whaling, land-based marine pollution, air pollution, ozone depletion, and climate change) that shows that science advice mechanisms function most effectively where a balance between scientific credibility and policy involvement has been achieved (Andresen, Skodvin, Underdal, and Wettestad, 2000). This conclusion derives from a broader empirical base (covering dumping, radioactive waste, fisheries, ozone depletion, land-based marine pollution, air pollution, satellite communication, nuclear nonproliferation, regional seas, trade in endangered species, whaling, and marine living resources) that shows the interactions between knowledge and institutional capacity to be the most important factor explaining effectiveness in environmental regimes (Underdal, 2002; Miles et al., 2002).

Chapter 4

Structure of Science Advice in the United Nations System Today

Introduction

The chapter aims at surveying and analyzing institutional arrangements for science advice in key organs of the United Nations. The functioning of the United Nations system is guided largely by considerations of geographical equity based on the five regions of the United Nations. Even where United Nations offices operate as subsidiary organs with a limited number of country representatives, the principle of geographical equity still applies. Existing procedures for providing advice are also guided by this equity principle. However, over the years, the United Nations has been increasingly expanding the use of science advice in decision making and introducing institutional adjustments that seek to balance scientific integrity and interaction between policy and science. This chapter analyzes trends by reviewing activities in the Office of the United Nations Secretary-General, functional commissions, programs, conventions, specialized agencies, processes and conferences, allied activities, and non-state actors. The focus of this chapter is to identify efforts made in the various organs to improve the use of science advice for decision making.

Office of the United Nations Secretary-General

The Office of the Secretary-General plays an important role in setting the tone for the functioning of the United Nations system. The Charter of the United Nations designates the Secretary-General as the "chief administrative officer" of the organization whose duties include carrying out activities entrusted to him or her by the Security Council, General Assembly, Economic and Social Council, and other United Nations organs. The Charter also empowers the Secretary-General to "bring to the attention of the Security Council any matter which in his opinion may threaten the maintenance of international peace and security." Carrying out this mandate involves keeping abreast of the latest international developments and working with member states to respond to the emerging challenges.

Many of the reports that the Secretary-General prepares for the various organs to which he is accountable require considerable scientific and technical input. For example, in April 2000, in preparation for the September 2000 Millennium Summit, the largest-ever gathering of heads of State or Government, the Secretary-General issued a report entitled, *We the Peoples: The Role of the United Nations in the 21st Century*. This was the most comprehensive projection of the UN's mission since its inception and called on governments to commit themselves to a 15-year program of work that addresses issues such as poverty, environment, conflict, and violence. The report dealt with policy issues such as information technology, biotechnology, and pharmaceuticals. The summit adopted the Millennium Declaration that set out a series of development goals that cannot be met without extensive scientific and technological input. Determining how to meet these goals will require knowledge of trends in science and technology.

The Secretary-General regularly convenes, at the request of governments, major international conferences and summits that are aimed at guiding international action on emerging issues. Many of these summits deal with issues that cannot be adequately addressed with effective science advice. The preparatory process for WSSD offers an illustration of the complex role of science and technology in global governance and the need for systematic institutional arrangements for science advice. Agenda 21 articulates the role of science and technology in two distinct ways. First, it recognizes the important role played by the scientific and technological community as a major stakeholder (as outlined in Chapter 31 of Agenda 21). The International Council for Science coordinated the input of this community into the preparatory process for WSSD in cooperation with the World Federation of Engineering Organizations (WFEO) and the Third World Academy of Sciences (TWAS), the InterAcademy Panel (IAP), and the International Social Science Council (ISSC). The contributions of the scientific and technological community were directed at government negotiators through submissions to the preparatory meetings.

Second, Agenda 21 recognized science and technology as being one of the means for implementing sustainable development (together with finance, human resources, and capacity building). This operational aspect of science and technology has not received as much attention as other themes. It is notable that the United Nations Secretary-General's choice of five priority areas—water, energy, health, agriculture, and biodiversity—was partly influenced by the availability of a large body of scientific and technical knowledge in those fields (Juma, 2002b). Assessments of the role of science and technology in the implementation of sustainable development goals in these five areas can play an important role in identifying opportunities for actions. The existence of science advice capacity in United Nations organs that convene summits and major conferences would help in determining the need for and modalities for such assessments.

Furthermore, the Secretary-General is increasingly being requested by the Security Council to address new threats to international security, such as health, whose effective management requires access to the best available scientific and technical knowledge. Other emerging science-based issues that will require the involvement of the Secretary-General include environmental management. This prognosis suggests that enhancing the capacity of the Office of the Secretary-General to serve governments will involve greater reliance on scientific and technical information (Juma, 2000). This is particularly important, as the role of science in international governance is becoming an explicit part of international diplomacy (Juma, 2002a).

The Secretary-General has recently introduced a number of measures aimed at strengthening the functioning of his office. The appointment of a Deputy Secretary-General provides managerial backstopping and allows the Secretary-General to focus on diplomatic and other functions. Additionally, the Secretary-General has established an office of strategic planning that helps to mobilize external knowledge for agenda setting. *The creation of capacity for strategic planning in the office of the Secretary-General is a natural starting point to consider the role of scientific and technical information in the world's highest diplomatic office.*

Science advice may also play a role in helping the Secretary-General bring policy consensus to the various activities of the United Nations that fall directly under his purview. This opportunity is provided by his role as chair of the United Nations System Chief Executives Board for Coordination (CEB), the successor body to the Administrative Committee on Coordination (ACC). The CEB brings together the executive heads of 27 member organizations, including UN funds, programs, and specialized agencies, the WTO, and the Bretton Woods institutions (including the World Bank and International Monetary Fund). The functioning of this body could be supported by science advice on emerging and persistent issues of global importance. Such advice could provide a basis for collective international responses by the UN system.

The office of the Secretary-General and its analogues in other international organizations plays a central role in supporting overall governance of international affairs. As shown below, a science advisor has previously served the Secretary-General. This office was abolished in favor of a more representative committee structure. *Although a committee structure may serve the needs of individual government delegations, it is not a good substitute for strengthened capacity in the office of the Secretary-General to use science advice in decision making. Building such capacity as part of strategic thinking will enable the Secretary-General to better support the various deliberative bodies.*

Functional commissions

The United Nations, through the Economic and Social Council (ECOSOC), has set up regional commissions to carry out social and development activities in Africa, Europe, Latin America and the Caribbean, Asia and the Pacific, and Western Asia. There are also functional commissions dealing with human rights, narcotic drugs, crime prevention, science and technology, the status of women, sustainable development, population, and statistics. The Commission on Sustainable Development (CSD) and the Committee on Science and Technology for Development (CSTD) are the most relevant for purposes of this study. These commissions rely on a wide variety of advisory inputs, most of which are provided through consultancy reports that are not subjected to any systematic procedures or reviews. In addition to these categories of commissions, the United Nations also operates commissions whose activities relate directly to the management of natural resources.

Established in 1992 as a subsidiary body of ECOSOC, the CSTD is mandated to provide the UN with reliable analysis and policy recommendations and to design effective implementation measures on relevant S&T issues. Certain Commission functions were revised in 1998, as ECOSOC reviewed the membership, focus, and operations of the CSTD (Juma, 2002a).

Through current CSTD work, S&T issues of concern to the UN are studied, particularly related to (a) the role of the science and technology in development; (b) science and technology policy, especially in respect to developing countries; and (c) science and technology matters within the United Nations system. Thirty-three member states are elected by ECOSOC to the CSTD for a period of four years: eight members from Africa; seven from Asia; six from Latin America and the Caribbean; four from Eastern Europe; and eight from Western Europe, North America, and other related states. Nominated by these governments to the Commission are experts with the necessary knowledge and qualifications. Commission priorities are those most central to the United Nations thus far, such as information technology and biotechnology (Juma, 2002a).

At each session, the Commission elects a Bureau (a chairperson and four vice-chairpersons) for the next session. A Bureau elected each session by the Commission manages all activities between sessions, establishing ad hoc panels or working groups to analyze substantive issues chosen for each intersessional period. These panels and groups are nominated and invited by Commission members to take responsibility for the reports, prepared by the United Nations Conference on Trade and Development (UNCTAD) secretariat and presented to the Commission at its regular session (Juma, 2002a).

Although the CSTD has undertaken a number of activities aimed at advising developing countries on issues related to information technology and biotechnology, it is too early to judge the impact of the efforts. Equally important is the long-term role of the CSTD in influencing the agenda of its host institution, UNCTAD, or being able to contribute to the functioning of other United Nations commissions, especially the CSD. *Governments continue to be interested in finding ways of enhancing the effectiveness of the Commission, especially in light of the fact that it is the only organ of ECOSOC that is charged with explicit responsibilities for advice on issues related to science and technology.*

Responsibility for monitoring the implementation of Agenda 21 rests with the CSD. More specifically, the CSD was established in 1993 as a functional commission of ECOSOC to: (a) monitor progress in the implementation of Agenda 21 and activities related to the integration of environmental and developmental goals throughout the United Nations system through analysis and evaluation of reports from all relevant organs, organizations, programs, and institutions of the United Nations system dealing with various issues of environment and development, including those related to finance; (b) consider information provided by governments, for example, in the form of periodic communications or national reports regarding the activities they undertake to implement Agenda 21, the problems they face, such as problems related to financial resources and technology transfer, and other environment and development issues they find relevant; and (c) review the progress in the implementation of the commitments set forth in Agenda 21, including those related to the provision of financial resources and transfer of technology.

The CSD has served largely as a continuation of UNCED, and much of its work was carried out through diplomatic negotiations (Wagner, 1999; Chasek, 2001a). Little effort was made to review the implications of the mandate—especially monitoring functions—for the secretariat that supports the CSD. The CSD has been effective as a body that brings governments together to pursue negotiations on specific aspects of Agenda 21. But this function has dominated its work, and little effort has gone into using scientific input to monitor the implementation of Agenda 21. As part of the preparations for the World

Summit on Sustainable Development (WSSD) the scientific community, operating through ICSU, has recommended that the CSD establish an office of science advice. This recommendation is in part based on the view that the CSD cannot effectively implement its monitoring role without building capacity for addressing scientific and technological issues.

Both commissions have been undergoing internal reform. First, governments have been concerned about the effectiveness of the CSD in its role as an organ that monitors the implementation of Agenda 21. The main innovation in the functioning of the CSD, however, has been the introduction of stakeholder dialogues aimed at enriching intergovernmental negotiations with additional information provided by non-state actors (Correll, 1999b). The focus for reform has therefore been on presentation and broader participation of the various stakeholders in intergovernmental processes. Little attention has been paid to improving the quality of information used in deliberations. Reforms in the CSTD, on the other hand, have focused on improving the functioning of its expert groups to improve the quality of the information provided. However, despite these reforms, much of the analytical work is still carried out by the secretariat. Efforts to engage the scientific community are still nascent and are implemented largely through workshops. Future reforms in bodies might involve increasing interactions between science and diplomacy in the CSTD and enhancing the technical credibility of information provided for decision making in the CSD.

Programs

Much of the operational work of the United Nations is carried out through programs with activities in specified sectoral or multisectoral areas, such as trade, development, refugees, children, drug control, volunteers, food, environment, population, relief, and human settlement. The UNEP is one of the most science-intensive organs in the system, although a large proportion of its work is devoted to conference diplomacy. The organization has played a leading role in linking science to diplomacy, and a large number of institutional innovations in this field have been inspired or nurtured by the organization. It deals with all aspects of the environment, including water. The UNEP's principal objective is to provide information and assessments to the broad international community and ultimately to the CSD for aggregation and policy decisions. A great portion of UNEP's work is scientific and technical in nature, intended to generate knowledge on environmental management and sustainable development.

UNEP acts as the convener for a number of scientific advisory groups, such as the Ecosystem Conservation Group (ECG), the Scientific and Technical Advisory Panel of the Global Environment Facility (GEF), the Intergovernmental Panel on Climate Change (IPCC), and the Joint Group of Experts on the Scientific Aspects of Marine Environment Protection (GESAMP).

> **Box 4.1: Science advice for marine resources conservation**
>
> Sound decisions in the field of marine conservation depend on the availability of high-quality and reliable scientific information. To be useful in global and regional programs, this information must be gathered from experts from all corners of the globe, and it must be collected and presented in a consistent way. This is why in 1969 the UN system created GESAMP, made up of scientists from all regions who serve in their individual capacities. Its purpose is to: (1) respond to requests for advice on specific scientific questions submitted by sponsoring organizations; (2) prepare periodic reviews of the state of the marine environment regarding marine pollution; and (3) identify problems that require special attention or programs. The regular sessions of GESAMP have produced a large number of reports and studies relating to marine pollution problems, covering a wide range of topics of relevance to marine environmental policy over more than 30 years.
>
> Working groups of experts selected by the organizations sponsoring GESAMP carry out the work of GESAMP. More than 750 experts have participated in the work of GESAMP's 32 working groups. The coordination of GESAMP is provided by a joint secretariat of the sponsoring organizations. At present, UNEP is the sole sponsor of all nine presently active GESAMP working groups and subgroups. Recently, a GESAMP subcommittee raised the issue of the lack of a global body on oceans to receive their advice and proposed a new informal consultative process in the General Assembly. The committee claimed that concerns about GESAMP's effectiveness were not due to GESAMP's own working methods but rather to the lack of a consolidated cross-sectoral process to carry forward its advice to the intergovernmental level.

UNEP also has historically worked closely with specialized agencies of the United Nations and other technical bodies by utilizing its convening power on environmental issues. One of UNEP's most enduring contributions to global environmental governance has been its record in mobilizing scientific and technical knowledge to support international environmental norm setting. These efforts have generally culminated in conventions, action plans and strategies, research agendas, and political declarations. In addition, science advice mobilized through UNEP has been used to support decisions in other bodies dealing with environment-related issues (Tolba, 1998).

Science advice in UNEP is provided through a variety of arrangements depending on specific needs. The most dominant mode has been the use of expert committees and groups that are convened by the Executive Director, who in turn serves as a knowledge broker linking science to policy. Acting within the limits of the mandate provided to it at the 1972 Stockholm United Nations Conference on the Human Environment, UNEP made effective use of its agenda-setting responsibilities to bring science to bear on international environmental policy (Gosovic, 1992; Tolba, 1998). *The quality of leadership, especially in its capacity to manage the policy and science worlds, played an important role in the capacity of the organization to serve as a knowledge broker* (Cheyes and Cheyes, 1995). UNEP successes in areas such as environmental protection of regional seas can be traced to its capacity to bring scientific and technical knowledge to bear on decision making and action.

The quality of the leadership would not have expressed itself effectively without the support of a competent secretariat that provided continuity and administrative support for the organization and member states. Indeed, UNEP has over the years emerged as the secretariat of several international conventions dealing with issues such as biological diversity, ozone depletion, persistent organic pollutants, chemical pollution, hazardous waste, migratory species, and trade in endangered species. In addition, UNEP hosts a number of scientific assessments dealing with themes such as international waters and ecosystems. The provision of science advice for international policy making thus has coevolved with the growth and strengthening of secretariat functions.

Programs of the UN play a key role in implementing operational action plans. This is done directly or though partnerships. In some cases, hybrid institutions such as the GEF have been created. The GEF was established in 1991 to provide financial assistance to meet the incremental costs associated with the implementation of global environmental commitments. It operates with UNDP, UNEP, and the World Bank as its implementing agencies. The GEF, which has a triennial budget of over $2 billion, promotes international cooperation, supports actions to protect the global environment, and provides funds to developing countries and those with economies in transition for projects and activities in one or more of four focal areas: biological diversity, climate change, international waters, and the ozone layer. It has recently added desertification and control of chemical pollution to its funding priorities. Science advice to the GEF is provided through the Scientific and Technical Advisory Panel (STAP) that is administered by UNEP. UNEP provides the STAP Secretariat and performs liaison functions between the Facility and STAP. The Panel is comprised of 12 persons appointed by the executive director of UNEP in consultation with UNDP, the World Bank, and the GEF Secretariat.

The STAP mandate, as approved by the GEF Council in October 1995, includes: (a) strategic advice as a means to advance a better understanding of issues of the global environment and how to address them; (b) the development and maintenance of a roster of experts; (c) elective review of projects; (d) cooperation and coordination with the scientific and technical bodies of conventions; and (e) providing a forum for integrating science and technology and acting as link between the GEF and the wider scientific and technical community.

STAP has so far carried out a variety of assessments that range from reviewing individual projects and providing advice on the extent to which they conform to the mandate of the GEF to providing general advice to the organization on how to deal with new issues such as support for research-related activities. Its visibility and impact, however, have been overshadowed by the activities of the implementing agencies of the GEF, as well as the GEF secretariat, with their own dynamics and methods for securing science advice. The GEF remains one of the most important sources of dedicated funding for sustainable development activities in the fields of water, energy, fisheries, and oceans through its support for activities in the fields of biodiversity, climate change, international waters, and land degradation.

STAP has on the whole not fully realized its potential for a variety of reasons. First, all its members are appointed and retire at the same time. This practice does not foster continuity, and as a result STAP has not been able to foster a collective identity. Second, it does not interact on a regular basis with the GEF secretariat, which is a key reference point for its operations. Third, although it maintains its autonomy, it does not interact actively with the political process except through occasional participation in meetings. Finally, STAP has had little opportunity to embark on major studies aimed at giving long-term strategic advice to the GEF beyond the constraints of existing operational priorities. This limitation may be linked to the fact that the GEF itself is reconstituted every three years. Such short time horizons would serve as a disincentive for long-term strategic thinking (GEF, 2002).

STAP could play an important role in shaping the direction of support for these activities and strengthening scientific contributions to the GEF, but doing this will require some changes in the mandate, procedures, and staffing of STAP to enable it to make full use of the experts on its roster. It will take the strengthening of its secretariat, whose executive secretary could function as the *de facto* science advisor to the GEF, with more direct and regular contacts with the GEF secretariat itself. Furthermore, its studies, especially on major areas that affect the allocation of resources in the organization, would need to be undertaken in a manner that gives them the requisite legitimacy, credibility, and relevance to emerging international concerns.

On-going discussions on the need to reform STAP illustrate two important trends. First, they address issues related to the credibility of the science advice provided. For example, STAP maintains a roster of experts who are in theory called upon to review GEF documents. However, the roster has not been fully utilized. This concern is not unique to STAP. Rosters have been set up under other bodies, especially conventions, with similar capacity utilization concerns. Equally critical are issues related to the effective use of scientific and technical information generated by STAP. This can be enhanced by more systematic interactions between STAP and the policy-making organs of the GEF. Enhancing scientific credibility and facilitating interactions between science and policy will require strengthening the STAP secretariat and formulating rules that reflect the new challenges.

Conventions

There are over 200 regional and international conventions that deal with the environment and sustainable development. These conventions provide a wealth of accumulated experience on the relationships between science and international politics (Bolin, 1994; Underdal, 2000). This experience shows that successful cases of science advice entail institutional arrangements that grant intellectual autonomy on the one hand and allow interactions between science and policy making on the other (Skodvin and Underdal, 2000). Achieving this requires careful consideration of the institutional arrangements for science advice as well as other relevant factors like leadership and knowledge brokerage capabilities (Skodvin, 2000).

Conventions are an interesting institutional arrangement because they seek to be internally consistent and coherent. They express this through the design of their internal organs to reflect their specific functions. Having all the necessary subsidiary organs under one institutional structure helps to facilitate feedback and internal learning. The continuing efforts to improve science advice in the various conventions are an illustration of this social learning process (The Social Learning Group, 2001a; The Social Learning Group, 2001b). There is also considerable transfer of lessons among the conventions, and experiences gained under one treaty are becoming the basis for improvements in others. Government delegates and secretariat staff seeking consistency among the functions of the various international institutions often affect the knowledge transfer. Other sources of procedural consistency include the use of United Nations rules of procedure (Kaufman, 1996; Boyer, 2000).

Most of these conventions were negotiated on the basis of evidence of environmental degradation provided through scientific research. These conventions fall into broad categories covering issues such as atmosphere (including energy use), biological diversity (including forests, aquatic, and marine life), oceans and marine resources, land use change (including desertification and freshwater resources), and chemicals.

Most of the conventions deal with specific environmental issues; examples are the Basel Convention on the Control of Transboundary Movements of Hazardous Wastes, the Convention on the International Trade in Endangered Species of Wild Fauna and Flora (CITES), the Convention on Migratory Species (CMS), and the Ramsar Convention on Wetlands. Others, however, are broad in scope and deal with sustainable development issues. These include the United Nations Framework Convention on Climate Change (UNFCCC), the Convention on Biological Diversity (CBD), and the United Nations Convention to Combat Desertification (UNCCD). Most of the global environmental and sustainable development conventions were negotiated and ratified in the last three decades.

The conventions use a variety of science advice approaches. CITES and Ramsar, for example, rely on small expert groups for science and technology advice. Each of these groups consists of a small number of experts selected on the basis of regional representation. The groups are responsible for identifying and facilitating the research for scientific and technological questions that arise in connection with the specific issue. In contrast, the Basel Convention, CBD, CCD, and UNFCCC all rely on open advisory bodies for science advice. In this scenario, all parties are free to appoint experts from different sectors. These bodies typically convene working groups and *ad hoc* expert committees to fulfill the identified science and technology advisory goals. CBD and UNFCCC also have established rosters of experts to assist the open advisory bodies in their tasks (Fritz, 2000).

The establishment of the IPCC in 1988 marked a significant step in defining a science and technology advisory mechanism to assess the issues related to climate change. The creation of this panel was an example of foresight on the part of the World Meteorological Organization and UNEP. It was created before the UNFCCC was completed and adopted. The first assessment report of the IPCC was one of the main resources for the designers of the UNFCCC (Agrawala, 1998a).

The IPCC is notable for its semi-autonomous nature within the greater bounds of the UN at large. Its main body, the IPCC Bureau, is comprised of representatives of member governments. This group is responsible for choosing the IPCC chair, a technical expert in a field related to climate change. The Chair assists the Bureau in devising the IPCC initiatives and assigning the designated investigations to a set of working groups. The working groups and task forces that represent the other main part of the IPCC are divided by theme. Each working group is responsible for researching and reporting on the set of inquiries handed down from the Bureau. The members of the working groups are experts from a wide range of disciplines, including earth and atmospheric sciences, technology, economics, and social sciences, and are chosen by members of the IPCC Bureau. To accomplish the tasks assigned to them, members of the working groups evaluate existing data, initiate new research, and consult with *ad hoc* commissions. Their findings are then returned to the IPCC Bureau for review and eventual dissemination (Agrawala, 1998b).

The IPCC produces reports that address the scientific understanding of climate change, environmental, economic, and social impacts, and possible mitigation steps. Reports are in the form of assessments, technical reviews, and special reports. The IPCC also publishes its methodologies and supporting material. The IPCC has been responsible for three assessment reports since its inception. The third, published in 2001, included segments on the scientific basis, potential impact, regional vulnerability, and mitigation priorities for climate change. Examples of IPCC special reports are: *Aviation and the Global Atmosphere*, published in 1999, and *Emissions Scenarios*, published in 2000. Technical papers have included *Implications of Proposed CO_2 Emissions Limitations*, released in 1997, and *Technologies, Policies and Measures for Mitigating Climate Change*, released in 1996.

Box 4.2: Review process for IPCC reports

The production and review of IPCC reports includes the following steps:

Governments provide Bureau members and collectively decide on cochairs. The process of choosing individuals for other functions (working groups, authors, reviewers, etc.) is overseen by the governments, via the Bureau.

The working groups, the members of which rely on experts and ad hoc committees for the most up-to-date and accurate information, write a report.

The report is then sent to substantive editors who review the content and accuracy of the working groups' findings.

The edited reports are sent out to experts in the field for peer review.

The peer-reviewed reports are sent to each country on the IPCC Bureau for individual review.

After this review process, the report is sent to the Bureau for a final review and acceptance or rejection.

The IPCC Secretariat publishes and disseminates the final product for general consumption.

Over time, the IPCC has shifted its focus slightly to include matters of economics and the social ramifications of climate change, in addition to the scientific causes and effects. There has been a genuine effort on the part of the IPCC chairs and Bureaus to take a multidisciplinary approach to investigating climate change. They are including more stakeholders in their activities, with the inclusion of various nongovernmental organizations (NGOs) in their research and review processes. The IPCC has attempted to become more diverse, as well, and has made many efforts to ensure equal representation from northern and southern countries, including the election of two coordinating lead authors for each report chapter–one from the North and one from the South. Periodic reassessments and revisions are key activities in the effective maintenance of the IPCC.

IPCC stands out as a remarkable innovation in science advice in the United Nations system. However, a number of concerns continue to be voiced, especially by developing countries. The IPCC illustrates the strengths and difficulties of integration of science advice with policy making. First, the fact that IPCC studies rely on peer-reviewed research for assessments discriminates against scientific input from countries whose scientists do not have opportunities to release their studies in peer-reviewed journals. Similar imbalances exist in regard to the distribution of research activities in general, and this has prompted interest in helping to build research capacity in developing countries.

The IPCC processes that allow for interactions between scientists and policy makers help to reduce these concerns, but the pressure to help strengthen research capacity in developing countries continues to be a central theme in IPCC activities. These concerns are not unique to IPCC and continue to bedevil fields such as biological diversity research. For example, Diversitas, ICSU's research program on biological diversity, is built largely on research networks in industrialized countries with the hope that over time the program will expand to include more developing country teams. But despite these concerns, IPCC's role as a leading example of science advice in the United Nations system remains unparalleled.

The importance of ensuring that science advice is conducted through the appropriate institutional arrangements is illustrated by the case of the *Global Biodiversity Assessment*. In 1995, the steering committee of the Global Biodiversity Assessment (GBA) released its report together with a summary for policy makers. The 1,140-page study was hailed as the most comprehensive analysis of the science of biological diversity ever carried out, the culmination of two years of research and data collection (Heywood and Watson, 1995). Funded by the GEF and UNEP, the independent assessment was the product of over 1,500 scientists and experts from all parts of the world, including over 300 authors. The text of the assessment was peer reviewed for scientific accuracy in an elaborate process that was unprecedented in the biodiversity community. The final product was a comprehensive assessment of the state of knowledge of the characteristics, magnitude, distribution, and function of biodiversity as well as the costs and benefits of anthropogenic influences. It was intended to serve as a basis for science advice for the CBD in a manner analogous to the role of IPCC reports in the UNFCCC process.

But despite the effort that went into preparing the report and ensuring that the best available scientific knowledge was utilized, the CBD did not receive it with enthusiasm and declined to offer an outright endorsement. The parties to the CBD deemed the report to be an externally generated document that they had not expressly asked for. This concern, however, masked critical issues related to the interactions between science and policy. The GBA process was initiated before the CBD came into force, and as a result there was no interaction between the scientific community that prepared the report and the policy making body of the CBD.

Policy makers were particularly concerned about undue influence over the future of the CBD through the work of the GBA. They were also concerned that the creation of the GBA was likely to undermine the functioning of the SBSTTA. The concerns were reinforced by the perception that the GBA reflected the conservation agenda articulated by industrialized countries and was not in keeping with the overall spirit of the CBD. "The overall framing of the assessment centers on the convention's first objective, conservation of biological diversity. The second object, sustainable use of its components, is discussed, yet annexed to the conservation theme. The convention's third objective, however—the 'fair and equitable sharing of the benefits' of the use of biodiversity—is almost entirely ignored" (Biermann, 2002, p. 208).

The GBA's lack of impact arose largely from the fact that the process of building scientific consensus, which is equally critical in the field of biological diversity, did not benefit from interactions with the policy-making organs of the CBD. This is not to say that such interactions would have automatically resulted in the acceptance of the report. What this case shows is that there is a need for proximity between the process of building scientific consensus on the one hand and the policy context on the other. The IPCC process is different from the GBA because it predated the UNFCCC. As shown elsewhere in this report, there is considerable interaction between the two processes. The story of the GBA raises fundamental questions about future independent assessments such as the Millennium Ecosystem Assessment that have tacit support from the governing bodies of various conventions but are subject to uncertain procedures on how the assessment and political processes interact, especially on issues that are dominated by geopolitical differences between the industrialized and developing countries (Biermann, 2000).

The UNFCCC, CBD, and UNCCD have analogous internal science advice organs and follow similar procedures. The main difference between them is that the UNFCCC relies considerably on independent input provided by IPCC. The other two do not enjoy such support, and the relevance of establishing IPCC-like bodies continues to be a question of considerable debate. An examination of science advice under the CBD illustrates the opportunities and challenges in sustainable development conventions (Juma and Henne, 1997). Article 25 of the CBD established SBSTTA to: (a) provide scientific and technical assessments of the status of biological diversity; (b) prepare scientific and technical assessments of the effects of types of measures taken in accordance with the provisions of this Convention; (c) identify innovative, efficient, state-of-the-art technologies and know-how relating to the conservation and sustainable use of biological diversity and advise on the ways and means of promoting development and transferring such technologies; (d) provide advice on scientific programs and international cooperation in research and development related to conservation and sustainable use of biological diversity; and (e) respond to scientific, technical, technological, and methodological questions that the Conference of the Parties (COP) and its subsidiary bodies may put to the body.

On a regular basis, this body reports to the COP for CBD through a chair who is elected by the COP on the basis of geographical rotation. A Bureau of 10 country representatives similarly elected supports the chair. The COP designates the goals and guidelines for the SBSTTA. All of the rules and regulations that guide the work of the SBSTTA are outlined in a comprehensive *modus operandi,* developed on the basis of experiences in other United Nations bodies. The SBSTTA often calls on experts in the field as well as ad hoc groups to respond to questions that are posed to it by the COP. The Secretariat of the CBC maintains a roster of experts designated by governments on a variety of topics. There are now 17 rosters of experts from which the Secretariat can draw when constituting *ad hoc* panels. The *modus operandi* also provides for the creation of liaison groups that can work directly with the executive secretary in developing documentation for SBSTTA meetings. The *modus operandi* takes into account the role of nongovernmental organizations and other relevant bodies in providing scientific and technical input.

The SBSTTA is an open-ended body whose membership is similar to that of the governing body, the Conference of the Parties (COP). The COP decides on what kind of science advice it requires. This is then presented as papers prepared by the staff of the Secretariat and in some cases with input from governmental and nongovernmental experts who operate through liaison groups established by the Secretariat. The papers are considered by the SBSTTA, and its recommendations are presented to the COP for consideration. In addition, the SBSTTA also hears presentations by independent experts. Although the CBD has increased the participation of the scientific community in its deliberations over time, there is room for improvement of the functioning of the science advice processes. For example, there is no distinction between documents prepared to elaborate on routine issues that are already part of the work program and those on areas of major international concern that demand a broader base for scientific input. Such issues could be the subject of more systematic consideration through assessments that could be organized along the lines of the functioning of the IPCCC.

However, pursuing this approach would require changes in the way the convention functions. For example, it would require putting in place procedures that are more elaborate than those used at the moment. Second, such an approach may entail reliance on outside bodies through a process of delegation that is consistent with the *modus operandi* of the SBSTTA. In fact, the SBSTTA already relies to a large extent on input from other bodies and could set guidelines for the way these organizations prepare their input to reflect the requirements of legitimacy, credibility, and salience. Such a move would entail strengthening the capacity of the office that deals with scientific and technological affairs in the secretariat.

On the whole, the CBD continues to make efforts to improve the functioning of its science advice system, but the structure has not allowed for a balance between scientific credibility and policy involvement. The convention processes have placed a higher premium on policy involvement than on the need to improve the scientific credibility of the reports on which advice is provided. The incremental measures adopted through revisions of the *modus operandi* have fallen short of the need to embark on major assessments on key areas such as marine and coastal biodiversity, forests, agriculture, and freshwater. This is in turn a consequence of political considerations among governments and their hesitation to subject to international scrutiny areas that they consider strategic. Similarly, political considerations contributed to the reluctance to undertake biosafety assessments as a basis for negotiating the Cartagena Protocol on Biosafety (which is discussed elsewhere in this report). Given such political sensitivities, it may be useful for the CBD to rely on external bodies under specific guidelines that promote credibility, legitimacy, and salience. The basis for such delegation already exists. For example, the convention's financial mechanism is implemented by the GEF.

Several other conventions are also involved in efforts to improve the functioning of their science advice organs and processes with a focus on seeking a balance between scientific credibility and policy involvement. For example, the Ramsar Convention has in recent years been improving the functioning of its Scientific and Technical Review Panel (STRP). One of the reforms aimed at promoting the autonomy of the panel was a decision of the contracting parties to the convention requiring STRP members to serve in their personal capacity and not as representatives of their countries of origin. The convention is also considering other measures that would seek to strengthen the review process under the STRP. The eighth meeting of the contracting parties scheduled for November 2002 will consider proposals that include detailed procedures aimed at making the STRP more credible, transparent, and efficient (Ramsar Bureau, 2002). The proposals recognize that improving the effectiveness of science advice is largely dependent on the nature and character of institutional arrangements designed to support the process.

Experience in the functioning of conventions shows the tension between the need to improve scientific credibility through independent processes that allow for peer review and the need to enhance interactions between science and policy. At the extreme level, government representatives selected to serve on scientific and technical bodies also participate on the governing bodies. This is a common phenomenon in open-ended bodies. This overlap may help to ensure that the recommendations of the advisory bodies are adopted by the governing bodies, a feature that may be seen as an indicator of effectiveness; however, it masks more serious problems related to the credibility of the science advice provided. Measures to address this concern include indicating that individuals cannot serve on both bodies or requiring that representatives to scientific and technical bodies serve in their personal capacity and not as officials of their governments. Such measures are complemented by strong peer review procedures.

Another important area of reform has been the recognition of indigenous or traditional knowledge as a source of input into science advice. The CBD, for example, explicitly recognizes the value of traditional knowledge in the conservation and sustainable use of biological diversity. Similarly, the CCD emphasizes the role of traditional knowledge as a key element in its operations. Also related to the broadening of the knowledge base is the role of non-state actors in international environmental governance (Correll, 1999). The participation of these groups, as well as the business community, has helped to widen the base for scientific and technical input. These efforts have also gone hand in hand with the strengthening of the secretariats of international organizations to support the participation of these groups.

The world's scientific expertise resides mainly in the industrialized countries, and the process of science advice is most advanced in these countries as well. *A call for the best known experts with experience on expert committees providing science advice in most fields is likely to find more candidates from the industrialized countries and few from developing nations. Such an imbalance could affect the legitimacy of science advice provided by such groups.* Indeed, there are a number of cases—for example, in the field of desertification control—where nongovernmental organizations with limited scientific and technical expertise had greater influence on negotiating processes than scientific committees (Correll, 1999a). In these cases the fact that the committees were made up largely of scientists from industrialized countries undermined the legitimacy of the outputs of the committees. Seeking to strengthen science advice at the international level will involve similar efforts at the national level in developing countries.

Promoting interactions between policy and science remains one of the key challenges in convention processes. Under the climate regime the science assessment process is independent of the convention. Interactions are facilitated through specific measures such as joint bureau meetings. Efforts to replicate this process in biodiversity-related conventions are still experimental. The MA is the most advanced effort to create science assessment mechanisms outside the convention process. It is too early to judge its impact. *On the whole, the majority of efforts to improve the functioning of science advice in convention processes focus on enhancing scientific credibility through broadening the input base and introducing or strengthening the peer review processes. These measures have been easier to achieve than the more complex challenge of promoting interactions between science and policy while maintaining scientific credibility.*

United Nations programs and funds represent some of the most influential global mechanisms for implementing sustainable development activities. The programs and funds deal with issues that require systematic use of the best available knowledge for decision making. The quality of decision making in their governing bodies can be strengthened through the provision of science advice. As shown above, many of the bodies are already exploring ways of strengthening their science advice activities. But doing so will require the establishment of internal mechanisms that coordinate science advice activities following the elements outlined in Chapter 2.

Specialized agencies

Much of the technical work of the United Nations is carried out through a number of specialized agencies dealing with issues such as labor, education, science, culture, food, agriculture, health, meteorology, telecommunications, postal services, intellectual property, development finance, civil aviation, and industrial development. Their functions are further divided into more specific areas, such as forestry and fisheries within agriculture.

The Food and Agriculture Organization (FAO) is the principal international organization concerned with fisheries, and it governs the agreements aimed at improved coordination in this regime. A specialized organization of the United Nations system, FAO has nurtured a large and competent fisheries department under an assistant director-general. Administration and program review is exerted through the director-general and the intergovernmental bodies, such as the Committee on Fisheries (COFI). FAO has established an Advisory Committee on Fisheries Research (ACFR) to deal with its internal research program and its relevance to support international fisheries objectives.

Box 4.3: Science advice for fisheries management

The Advisory Committee on Fisheries Research was first established in 1993 by the director-general of FAO. Its membership consists of scientists appointed in their own capacities by the director-general on the basis of their competence in the relevant disciplines and geographical distribution. Its main function is to study and to advise the director-general on the formulation and execution of FAO's work on all aspects of fisheries research. Special attention is given to the fisheries aspects of oceanographic research and to the impact of environmental change on the sustainability of fisheries. The Committee also acts as the advisory body to the Intergovernmental Oceanographic Commission of UNESCO on the fisheries aspects of oceanography.

Although the Committee is numerically small, consisting of only eight members, selection has been made to include the widest possible subject matter and geographical representation. The committee's mandate includes conservation and management of marine and inland fishery resources, increasing fish productivity through enhancement of wild resources and through aquaculture, improving the means of converting fishery resources into human food, and studying the dynamics of fishing communities and the socioeconomic consequences of government fishery policies. This is an ambitious menu for a small committee that meets only for several days once a year. However, ACFR is permitted to establish working groups, subject to approval of the Director-General and availability of funds. So far, only one working group has been at work.

There are also a number of science advisory bodies serving the seven or so regional fisheries organizations. An early example was the International Council for the Exploration of the Sea (ICES), established in 1902. A major part of the work of ICES is undertaken at the direct request of eight international regulatory commissions and national administrations for scientific information and advice on the conservation and management of fish and shellfish stocks, including the effects of pollution on the marine environment. This advisory function expanded significantly during the 1980s and has continued to do so. Today, ICES furnishes advice to international commissions, such as the North-East Atlantic Fisheries Commission (NEAFC), the International Baltic Sea Fishery Commission (IBSFC), and the North Atlantic Salmon Conservation Organization. In the fisheries field, ICES has its own advisory body, the Advisory Committee on Fishery Management (ACFM), which meets twice a year. Its findings and advice are supplied to member governments and fishery commissions and are subsequently published in the ICES Cooperative Research Report series.

Another area that is increasingly acquiring international attention is water. There are approximately 300 different agreements and treaties that relate to freshwater as a resource. A few of these have international applicability, such as the Convention of the Protection and Use of Transboundary Watercourses and International Lakes, the Ramsar

Convention involving the fundamental ecological functions of wetlands as regulators of water regimes and as habitats supporting a characteristic flora and fauna, and the Convention on the Law of the Non-navigational Uses of International Watercourses of 1997.

UNESCO's International Hydrological Programme (IHP) is the only science and education program of the United Nations in the field of hydrology and water resources. It is a vehicle through which Member States can upgrade their knowledge of the water cycle and thereby increase their capacity to better manage and develop their water resources. IHP is governed by an Intergovernmental Council, and its secretariat, housed at the Division of Water Sciences of UNESCO, Paris, plays a catalytic role in the execution of the Program. The UNESCO regional offices, five of which have hydrologists, play an increasingly important role in implementation. Regional cooperation is an important aspect of IHP's global program, and headquarters and the regional offices work closely with some 160 national committees and focal points in implementing its activities.

IHP was originally launched as the International Hydrological Decade (IHD) in 1965, and afterwards, as an international program. It has had a number of successive multiyear phases; the theme of the fifth phase, which extends from 1996 to 2001, is Hydrology and Water Resources Development in a Vulnerable Environment. IHP's objectives are to stimulate a stronger interrelation among scientific research, application, and education. Emphasis is on environmentally sound integrated water resources planning and management, supported by a scientifically proven methodology. IHP also sponsors post-graduate hydrology courses with emphasis on a multidisciplinary approach and prepares computer-based learning materials and, in cooperation with other UN agencies, a thesaurus of water resources terms.

> **Box 4.4: Science advice on freshwater**
>
> The Scientific Committee on Water Research (SCOWAR) was established by the 24th General Assembly of ICSU in 1993 to provide the objective scientific expertise in water resource problems required to address frontier issues in science. The purpose of the SCOWAR working group is to address the ecological consequences of altered water regimes. The specific objectives are:
>
> (1) To evaluate databases that relate to impact on freshwater biodiversity of nutrient cycling, productivity, invasion of alien species, and resistance to unusual disturbances;
>
> (2) To provide an overview of the ecological consequences of altered hydrological regimes; and
>
> (3) To suggest alternative scenarios for management using modeling approaches. The spatial scope is global, and the temporal scope focuses on the next 2-3 decades.
>
> The activities are coordinated with relevant ICSU bodies such as the Biospheric Aspects of the Hydrological Cycle (BAHC), the International Geosphere-Biosphere Programme (IGBP), and the Scientific Committee on Problems of the Environment (SCOPE) and with national and international bodies such as the National Center for Ecological Analysis and Synthesis (NCEAS) at the University of California at Santa Barbara and the International Hydrological Programme.

The Intergovernmental Oceanographic Commission (IOC), for example, was set up by UNESCO in 1960 to develop, recommend, and coordinate international programs for scientific investigation of the oceans and to provide ocean-related services to Member States. IOC is explicitly recognized as an international organization competent in marine scientific research within the context of the United Nations Convention on the Law of the Sea. While quasi-autonomous, the IOC is assigned to UNESCO for funding and staffing and is reviewed by the UNESCO General Conference and Executive Board and administered under authority of the UNESCO Director General.

The IOC Secretariat consists of some 20 professionals headed by an executive director who is an oceanographer. The three component sections deal with Ocean Sciences, Operational Observing Systems, and Ocean Services. The Secretariat has established many electronic tools to enable policy makers to have access to the scientific and policy world, including a database, the International Oceanographic Data and Information Exchange System (IODE), a roster of experts (GLODIR), and a website with Internet links to the world of oceanography. Over the years since its creation, IOC has had an active partnership with the Scientific Committee on Ocean Research of ICSU in organizing international scientific cooperation.

IOC collaborates with and exchanges reports and technical information with many other organizations within the United Nations system. An example of an important joint program is the Joint Commission on Oceans and Atmosphere (J-COMM), created in

1999 in actions by UNESCO/IOC and the World Meteorological Congress of WMO. The formation of a high-level intergovernmental body of experts in oceanography and meteorology represented a culmination of more than 30 years of IOC/WMO cooperation on ocean observing systems and services. The Commission addresses such problems as El Niño/La Niña prediction as well as more general studies of global climate and climate change.

The IOC's Global Ocean Observing System (GOOS) is one of the global observing systems set up after the United Nations Conference on Environment and Development. It also contributes to the UNEP Global Earth Monitoring System (GEMS) and is a part of United Nations' Earthwatch. It is managed with an international steering committee made up of representatives of IOC and the involved agencies, including ICSU/SCOR.

Specialized agencies of the United Nations are involved in a wide variety of science advice activities including the use of ad hoc technical groups to standing committees set up to provide advisory services. This is mainly because the mandates of these agencies require them to address scientific and technical issues on a regular basis. In addition to workshop reports and staff papers, these agencies rely on commissioned papers—mainly from independent consultants—to inform many of their internal activities. *These approaches, however, are not guided by systematic procedures designed to ensure that the input provided—especially on major issues—is credible, legitimate, and relevant. The growing impact of scientific and technical matters on deliberations and activities of these agencies will increase their pressure on the chief executives to find more systematic and coordinated ways of security science advice. This may entail the creation of offices specifically charged with such a mandate and the adoption of procedures that are designed to ensure legitimacy, credibility, and the relevance of the science advice provided.*

Processes, conferences, and joint research activities

In its norm setting and research activities, the United Nations operates through a variety of processes, conferences, programs, and forums. Processes are activities that are set up to function within a limited time frame. They are usually based on a single conference but have longer-term programs of work. Some of the processes evolve into more permanent forums. For example, the Intergovernmental Panel on Forests was transformed into the Intergovernmental Forum on Forests. Similar arrangements also exist for research programs that are jointly sponsored by two or more agencies.

Agenda setting on global issues often requires processes that involve a large number of stakeholders. Independent commissions have played an important role in defining the agenda for action in fields such as global governance, health research, disarmament and security, international development, population, food, water, solar energy, oceans, cultural diversity, large dams, sustainable development, and forests (Dowdeswell, 2001). The World Commission on Environment and Development (WCED), chaired by Norwegian Prime Minister Gro Harlem Brundtland, had a lasting impact on the global agenda. Its success can be attributed partly to the quality of input and the level of engagement with the public. The WCED established a strong knowledge base that included scientific assessments and subjected the various reports to external review. It reached out to a diversity of constituencies and relied on public hearings and testimonies for its work.

In addition to building on a strong scientific basis for its work and reaching out to the public, the Brundtland Commission had a strong basis in formal diplomacy by building on a resolution of the United Nations General Assembly. The impact of the engagement of the scientific community in the Brundtland Commission is also reflected in its outcomes. Its report, Our Common Future, provides a strong scientific and technological agenda for implementation of sustainable development goals (WCED, 1987).

The degree to which independent commissions rely on scientific and technical input varies considerably. Issues such as health, oceans, dams, and sustainable development have strong scientific underpinnings. The commissions have also helped to underscore the importance of broadening the knowledge base for development. For example, the World Commission on Culture and Development (WCCD, 1995) helped to underscore the role of traditional knowledge and cultural diversity in development.

The functioning of the World Commission on Dams—founded by the World Bank and the World Conservation Union (IUCN)—is an example of issues of global concern that can only be effectively addressed through a structured process that encourages the use of scientific and technical information (WCD, 2000). The commission relied on a "knowledge base" including case studies and other technical material and review papers that provided a synthesis of the state of the art. The peer-reviewed papers identified various scenarios for future consideration, developed decision tools based on best practices, and proposed criteria and guidelines for decision making where the construction of large dams was an option. These papers provided the basis upon which public consultations were conducted. WCD explicitly sought to ensure that the development of new standards for managing large dams was founded on the best available information that was then used in public deliberations. The case of WCD reveals the growing interest in ensuring that democratic practices of decision making on major global issues are guided by the best available knowledge.

Scientific networks serve an important role in providing input into international decision making. Whereas many of the networks are independent activities with no formal institutional affiliations, some are part of existing organizations. The Species Survival Commission (SSC) of IUCN is an example of a network of scientists operating under the auspices of an international organization and generating information that is used for decision making. SSC has 7,000 volunteer members working in all regions of the world. They include wildlife researchers, government officials, wildlife veterinarians, zoo employees, marine biologists, wildlife park managers, and experts on birds, mammals, fish, amphibians, reptiles, plants, and invertebrates, as well as experts on wildlife policy.

The members operate though more than 120 Specialist Groups and Task Forces. Some groups deal with conservation issues related to particular plant or animal groups, whereas others focus on topical issues such as sustainable use of species or reintroduction of species into their former habitats. The most important output of the SSC is the IUCN Red List of Threatened Species. The IUCN Red List is the world's most comprehensive inventory of the status of plants and animals worldwide. It uses a set of criteria to evaluate the extinction risk of species and subspecies. The criteria are: "extinct," "extinct in the wild," "critically endangered," "endangered," "vulnerable," "near threatened," and "least concern."

A species is listed as threatened if it falls in the critically endangered, endangered, or vulnerable categories. The IUCN Red List vividly conveys the urgency and scale of conservation challenges to the public and to policy makers and motivates the international community to take the necessary action. The SSC has over the years played a key role in keeping the public informed about trends in the conservation and use of biological diversity. The information is used for policy making within IUCN as well as in other international treaties and organizations dealing with biodiversity conservation, such as CITES, the Convention on Migratory Species (CMS), the Ramsar Convention on Wetlands, and the Convention on Biological Diversity (CBD). Placing species on the IUCN Red Data List is a complex process that involves interactions between the scientific community and political processes at the national and international levels.

Although intergovernmental processes, conferences, and joint research activities appear ephemeral, they often have major impact on global governance. United Nations summits, for example, have far-reaching implications for global governance and are often the source of significant follow-on activities. Many major conferences lead to the creation of new institutional arrangements. It is therefore important that the preparatory process for major conferences and summits benefit from science advice. This should be provided through the secretariats convening the events.

BOX 4.5: Some major international activities on freshwater involving science

United Nations System
- UN Commission on Sustainable Development
- UN System-Wide *World Water Development Report* (to be completed 2002)
- UN Inter-Agency Working Group on Water in Africa
- UN ACC Sub-Committee on Water Resources (Chaired by Director of IHP)
- For a comprehensive list of UN activities on freshwater see *Report of the Secretary-General on Activities of the Organizations of the UN System in the Field of Freshwater Resources* (E/CN.17/1998/3 of April/May 1998) (http://www.un.org/documents/ecosoc/cn17/1998/ecn171998-3.htm)
- For a list of UN water-related databases (as at 10 August 1999) see http://www.un.org/esa/sustdev/watbase.htm

UNEP
- Global International Waters Assessment (GIWA). For an annotated list of links, see: www.giwa.net
- Global Programme of Action for the Protection of the Marine Environment from Land-based Activities
- WCMC, *Freshwater Biodiversity: A Preliminary Global Assessment* (1998)
- Managing Water for African Cities (UNEP/UNCHS, UNF/UNFIP funded)
- UNEP Global Environment Monitoring System Freshwater Quality Programme (GEMS/WATER)

UN and Affiliated International Organizations
- FAO has numerous activities under its Water Resources, Development and Management Service.
- GEF Global Action on Water has funded projects in the areas of water scarcity, pollution reduction, preventing conflict, and land-based sources of marine pollution.
- WHO: Global Assessment 2000: Status of the Water Supply and Sanitation Sector; Divisions responsible for water projects include: Water, Sanitation and Health; Child Health and Development; and Control of Tropical Diseases. A *Report of the UN Secretary-General on Progress in Providing Safe Drinking Water Supply and Sanitation for All during the 1990s* is also to be published.
- Water Supply and Sanitation Collaborative Council (WSSCC) was established within WHO as follow-up to the UN International Drinking Water and Sanitation Decade (1981-1990). A Fifth Global Forum on water took place in November 2000 in Rio de Janeiro, Brazil
- UNESCO - International Hydrological Programme (IHP) has many activities including the new Hydrology for the Environment, Life and Policy (HELP) program and the Flow Regimes from International Experimental Network Data Sets (FRIEND).
- UNESCO - Intergovernmental Oceanographic Commission (IOC) - Global Directory of Marine and Freshwater Professionals (GLODIR)
- UNDP Sustainable Energy and Environment Division (SEED) works with the Global Water Partnership, hosts the GEF International Waters secretariat, and has many other activities. For an extensive list of water-related links see http://www.undp.org/seed/water
- UNDP, World Bank, and 15 bilateral donors have a Water and Sanitation Programme
- The World Bank has many water activities and sponsored a Water Supply and Sanitation Forum in April 2000
- WMO has numerous activities under the Hydrology and Water Resources Programme, and the World Climate Research Programme has a Global Energy and Water Cycle Experiment (GEWEX)

Other International Activities
- Global Water Partnership
- International Water and Sanitation Centre (The Netherlands, UNDP, UNICEF, WHO, WORLD BANK, WSSCC) - maintains the UNEP/UNCHS urban water website at http://www.un-urbanwater.net
- World Commission on Dams (IUCN and The World Bank) - see http://www.dams.org
- World Water Council has initiated the World Water Vision (managed by IHP) and the World Water Forum (guided by the World Commission on Water for the 21st Century)

Source: **Report on International Scientific Advisory Processes on the Environment and Sustainable Development**, Prepared for the UN System-Wide Earthwatch Coordination, UNEP, by Jan-Stefan Fritz

> **Box 4.6: Science advice on marine pollution**
>
> The Global Investigation of Pollution in the Marine Environment (GIPME) is an international cooperative research program focused on marine contamination and pollution. It was established in response to the recommendations of the United Nations Conference on the Human Environment held in Stockholm in 1972. GIPME is cosponsored by the Intergovernmental Oceanographic Commission, the United Nations Environment Program, and the International Maritime Organization (IMO).
>
> GIPME investigations focus primarily on the coastal zone and shelf seas. The program assesses the presence of contaminants and their effects on human health, marine ecosystems, and marine resources and amenities, both living and nonliving. A Scientific and Technical Committee that meets once every four years develops the program. The day-to-day work within GIPME is conducted by three Expert Scientific Groups: the Group of Experts for Methods, Standards and Intercalibration (GEMSI), the Group of Experts on the Effects of Pollutants (GEEP), and the Group of Experts on Standards and Reference Materials (GESREM). The Scientific and Technical Committee for GIPME establishes the long-term direction and strategy for the GIPME Program. Its published reports are the Comprehensive Plan for GIPME in 1976 and the GIPME Implementation Strategy in 1984.

Allied activities

Science advice is also provided through other international non-UN agencies that work on scientific and technical issues. The International Energy Agency (IEA) has a staff of professionals and experts seconded from member states of the Organization for Economic Cooperation and Development (OECD). It interfaces with energy information and analysis sources in its member nations to gather country statistics and information concerning energy programs, research, and technology. It holds workshops to survey and disseminate information on the state of knowledge in energy technical areas. The IEA maintains on-line databases and a directory of international partner organizations in member states.

The IEA performs national assessments and reviews of member states' energy policies with the help of visiting teams of experts. It participates in joint programs with the Economic Commission for Europe and UNEP and provides background reports to the UN Commission on Sustainable Development. IEA has been keenly involved in the issues of greenhouse gas reduction and energy use and has participated in UNFCCC deliberations. The Governing Board of IEA is assisted by standing groups and special advisory committees, which bring together energy specialists from member countries.

The Millennium Ecosystem Assessment is a nongovernmental affiliate of the United Nations seeking to provide advice on biodiversity to organs of the UN. Its objective is to evaluate the usefulness of ecosystem goods and services to human development by assessing the damage caused by the degradation of ecosystems in various regions and communities. The information gathered during the assessment is to be used to determine national and international policy regarding ecosystems and the environment. This elaborate and comprehensive assessment is the first of its kind, in that it will include analysis of the entire spectrum of ecosystem goods and services as well as all possible impacts on ecosystems. The focus of the assessment is unique also, as it takes the perspective of human gain and loss, rather than a purely environmental standpoint.

The original concept of the Millennium Ecosystem Assessment came from a meeting of members of The World Resources Institute (WRI), United Nations Environmental Programme, United Nations Development Programme, and the World Bank who were looking for a more comprehensive and realistic assessment of ecosystem goods and services than was available in 1998. Members of these organizations pushed the concept to reality over the next few years, and today the MA is under way. The first step in this process was to establish an Exploratory Steering Committee to develop the plan of action and generate institutional support for the MA.

The functional structure of the MA consists of a Board, a supporting Secretariat, and five working groups. The Board oversees the assessment process and designates the course of action and focus of the working groups. Members of the board include expert representatives from academia, business, and civil society, in order to promote a broadly based final product.

The working groups each have a different focus, which is based on the conceptual framework designed by the Board of the MA. Each group has two chairs and multinational representation. The working groups will base their work on existing assessments and data, when available, and will collaborate with related institutional sources to fill in information gaps. These sources will include international assessment bodies as well as individual experts. The MA has been endorsed by and is intricately linked with the UN system, and it also receives support from national governments, the World Bank, and numerous foundations and funds.

The case of the MA shows the potential for delegating scientific assessment tasks to other organizations. Additional input into the United Nations system is provided through the participation of international organizations in the decision making processes of other bodies. Governing bodies that delegate scientific assessment responsibilities to other entities should provide guidelines that seek to enhance the credibility of the products of such assessments.

Non-state actors

Nongovernmental organizations (NGOs) and individuals play a significant role as sources of science advice and influence in the United Nations systems (Willetts, 1996; Raustiala, 1997). This is done through a variety of ways that include official participation at conferences—especially by those accredited to ECOSOC, where they have official speaking rights—and direct lobbying. NGOs occupy a special place in the United Nations system, and their influence has grown significantly over the years, especially in the post-UNCED period. Their activities have been closely associated with the wave of democratization that has swept across the world in the last decade. There is considerable variability in the character and orientation of these organizations. A number of them are specialized in certain fields of scientific and technological research and have played a key role in contributing advice to the international system.

Established in 1957, the Scientific Committee on Ocean Research (SCOR) of the International Council for Science (ICSU) is one of the oldest of the interdisciplinary bodies of the nongovernmental ICSU. The recognition that the scientific problems of the oceans require a truly interdisciplinary approach was embodied in plans for the International Geophysical Year in the 1950s. SCOR's first major effort was to plan a coordinated, international attack on the least-studied ocean basin of all, the Indian Ocean, which resulted in the International Indian Ocean Experiment of the early 1960s.

For the next 30 years, the reputation of SCOR was largely based upon the successes of its scientific working groups. These small multinational groups of not more than 10 members each were established in response to proposals from the national committees for SCOR, other scientific organizations, or previous working groups. In general, they are designed to address fairly narrowly defined topics that can benefit from international attention. All working groups are expected to produce a final report, organize a workshop or symposium, or otherwise make a significant contribution to advancing understanding of the topic at hand within three or four years of being established.

One of the most significant aspects of recent trends in global governance has been the increasing role of non-state actors influencing international decision making. This development has also been associated with the growing demand for democratic decision making worldwide, globalization, and the prevalent use of information technology. International organizations have been searching for ways of incorporating the participation of non-state actors in their activities. As in other areas of decision making, science advice processes are increasingly providing space for the participation of these actors and as a result broadening the base for knowledge input and enhancing both the credibility and legitimacy of the results. The pressure to open up political space is also being extended to other previously excluded groups. For example, the demand for gender balance is being added to the agenda for reform in international science advice bodies. The role of non-state actors is therefore part of the growing demand for the democratization of global governance systems.

Conclusion

Attempts to improve the effectiveness of science advice have resulted in a wide range of experiments aimed at balancing between scientific credibility on the one hand and interactions between policy makers and science on the other hand. These institutional experiments demonstrate the growing interest of the UN in strengthening the role of science advice in decision making and therefore forming the basis for recommendations. Efforts to improve science advice in the UN system have focused on two main goals: enhancing scientific credibility and promoting interactions between science and policy. This is achieved through institutional adjustments of rules and procedures and organizational development, especially in secretariats responsible for managing science advice processes. Many of these adjustments and reforms are in their early stages of development and therefore require guidance by offices and individuals whose functions include determining when science advice is needed, creating procedures and systems that can enhance scientific credibility and promote science-policy interactions, and maintaining liaison with other organizations.

Chapter 5

Findings and Recommendations

Introduction

The previous two chapters assessed science advice activities in the United Nations system. The assessment reveals that many of the elements of science advice outlined in Chapter 2 are being practiced but not in a systematic and consistent manner. For example, some international organizations have offices of the Science Advisor or their equivalents but they lack the appropriate procedures that can be used to strengthen the credibility of their work.

There is sufficient evidence to suggest that scientific leadership or entrepreneurship has played a key role in promoting international environmental agreements. These entrepreneurial efforts did not necessarily result in the creation of durable institutional arrangements and procedures for science advice. Where there is no effective secretariat guidance, governments may not easily agree on when science advice is needed and such determination is subject to political negotiations. As shown in previous chapters, much of the emphasis in the United Nations has over the decades been on geographical representation with little consideration for scientific credibility. But this is starting to change, and international organizations are starting to seek a balance between the two.

The management of biases and conflict is one of the weakest aspects of current science advice practices in the United Nations system. Although a few organs such as IPCC have developed measures to deal with this, the use of consultancy reports, workshops, and other tools of conference diplomacy promote the commingling of political interests and scientific assessments to levels that undermine the credibility of the outputs. The absence of clear procedures dealing with science advice exposes United Nations staff to political influence that compromises the management of data.

Although the reports may be outstanding products, the dominant processes often cast a shadow of doubt over their scientific standing. Many organizations are instituting measures to improve review processes and are developing rosters of experts from which they select names, but there is still excessive dependence on a limited number of consultants who often have undue influence on the outlook of the organizations that they work for. This concern can be addressed through an effective review process and broadening of the sourcing of papers. *Despite these concerns, there are encouraging signs that the United Nations is making efforts to address these issues and to adopt measures that will maintain sovereign equity while at the same time promoting scientific credibility. Continuous management of this balance requires expertise in science advice as well as the creation of appropriate institutional arrangements such as the office of Science Advisor. Such an office would be responsible for designing the procedures needed for science advice as outlined in Chapter 2.*

Findings

The United Nations system operates through a wide range of organs including the General Assembly, commissions, programs, research institutes, agencies, treaty bodies, forums, and conferences. Science advisory mechanisms of one sort or another are found throughout the system. However, executive heads of UN agencies, including the United Nations Secretary-General, have no systematic mechanisms that govern science advice for their operations. This is particularly important because these officers are responsible for alerting governments on emerging issues in their areas of jurisdiction, although decision making is reserved for intergovernmental processes. The various organs of the United Nations are autonomous and respond to their governing bodies rather than to other UN organs.

Many of the advisory committees of the United Nations provide scientific input into decision making but have no specific procedures that ensure quality and balance. Treaty bodies dealing with a variety of environmental and sustainable development issues have started to establish subsidiary bodies to incorporate science advice in their functions. However, the advice provided by these bodies is usually framed in the context of negotiating positions. There are a few quasi-independent science assessment processes in the United Nations that provide status reports on global scientific problems and recommendations to governments. The best known of these is the IPCC, which serves as a model for other assessment processes.

The agencies use a wide range of approaches to provide science advice depending on the functions of the various organs. Many of them rely on science advisory committees, staff reports, and consultancy studies, and the credibility of such reports varies considerably. Much of the advice provided in the United Nations system is undertaken indirectly through various deliberative bodies or through subsidiary advisory bodies. IPCC offers one model for fairly independent science advisory bodies within the UN system, but, as an intergovernmental body, its results already reflect a degree of government input, and it is not entirely independent. The Millennium Ecosystem Assessment (MA) is an alternative approach designed to preserve the independence of the advisory process.

There are presently considerable efforts by several organizations to improve the role of science advice in United Nations activities. The focus of the efforts has been to find institutional arrangements to improve the balance between scientific credibility and the need to promote interactions between policy makers and the scientific community, many of them inspired by IPCC. The recommendations provided in this report aim to support current efforts to improve science advice in the system.

Recommendations

As discussed in Chapter 1 and 2, there has been significant progress during the last decades in incorporating science advice into the work of the deliberative bodies and the programs of the United Nations. To support that progress, we make five recommendations that could greatly increase the UN system's ability to obtain science advice on the wide range of issues that it must address in the decades ahead. Two recommendations address staffing for science advice and the processes of science advice. The third recommendation is addressed to member states and calls upon them to develop their own capability for generating and using science advice for their own activities. The fourth recommendation calls on the governing bodies and assemblies to make greater use of external organizations that can offer science advice in the manner prescribed in this report, and the fifth urges the organization to continue other forms of science-based policy processes such as scientific assessments. The recommendations are based on general principles and practices that reflect current best practice in science advice. They are generalized for the UN organization as a whole, and they will naturally require some specific tailoring for the various organs of the United Nations.

Staffing the UN System for Increased Utilization of Science Advice

As discussed in Chapter 2, knowledgeable science advisers are essential for the organization and interpretation of science advice. The adviser must know when advice is needed, identify sources of information that may exist, select and recruit individuals who can become part of a committee to analyze a problem, and understand and implement the principles of the science advisory process referred to in Chapter 2. The chief executives of the governing bodies of the UN system would be greatly assisted in their strategic and operational responsibilities by having such science advisers as a part of their senior executive staffs.

Recommendation 1: Governing bodies of the United Nations that have substantial responsibilities for implementing sustainable development programs should each create an Office of the Science Advisor or equivalent facility or organizational function appropriate to its mandate.

The science advisory function should be within the office of the Secretary-General, Director-General, or Executive Secretary of the organ or conference and should serve the governing body of the organization through the Secretariat. These bodies include the governing bodies of specialized agencies and the governing bodies of specially convened international meetings, such as the World Summit on the Information Society to be held in 2003 and 2005 in Geneva and Tunis, respectively.

The purpose of these facilities would be to assist in the identification of scientific elements of policy questions, generate balanced and independent science advice, and help to interpret the resulting findings and recommendations for use in the policy process. The function of this office would be to:

1. Assist the governing body and the secretariat to recognize policy issues that require or would benefit from science advice,

2. Help the governing body and the secretariat to formulate the scientific questions to be asked,

3. Carry out or commission from an external organization the science advice process to respond to the questions,

4. Assist the governing body and the executive to interpret the meaning and degree of uncertainty in the resulting report, and

5. Help the governing body and the secretariat to understand the possible policy implications of the science advice.

The Science Advisor who heads the office should be chosen according to the principles elaborated above. It is important to ensure that the work of such Science Advisory offices be carried out in conformity with the best standards of the scientific community and carry the credibility of good science.

Management of data and information also requires a well-trained staff to serve the study committees. Staff can carry out extensive reviews of the literature, commission analytical papers on particular aspects of the topic, arrange expert testimony to the committee, and convene workshops where a variety of experts may present data and express their views to members of the committee. It is important to ensure the competence and independence of the staff and adherence to clear principles of science advice.

Science Advice that draws on established, proven principles creates credible, transparent, and authoritative information for decision makers.

The experience of scientific academies and other organizations involved in independent science advice has led to the development and evolution of a set of procedures that characterize good science advice. These include careful attention to statement of the task, recruitment of a broadly expert study committee with a balance of disciplines and views, management of external inputs, production of a public report, and independent peer review. The scientific community accepts processes containing these elements as indicators of objective science advice. They can be carried out by a unit within an organization, or commissioned from an outside entity. We therefore propose

Recommendation 2: Each such science advisory facility or organizational mechanism should adopt an appropriate set of general procedures based on those described in this report, adapted to any special circumstances of the organization. These procedures should be widely publicized within the corresponding diplomatic and scientific communities.

The purpose of this recommendation is to ensure that the UN organizations move toward reasonable uniformity of scientific advisory procedures based upon the best practices of the world scientific community.

Member governments can also benefit from establishing procedures for science advice.

In order to have available a pool of scientific and technical experts to participate in scientific advisory studies, it will be necessary to strengthen science and technology capability in many member states, especially developing countries. Experts from those countries will be more effective in the scientific advisory process and more influential in UN forums if they are experienced in the principle and practice of science advice. A program of support to develop science advice capability in the less developed member states will benefit the UN by ensuring wider geographical distribution among experts as it helps the governments to gain science advice for public policy making. The UN should also assist national science academies or other designated organizations within the countries, including non-state actors, to carry out advisory functions guided by the elements set out in Chapter 2. Suitable programs should include providing training and internships for staff, funding joint studies with other advisory groups or science academies in other countries, and recruiting experts associated with these organizations for UN science advisory committees.

Recommendation 3. The United Nations should help member states to strengthen their own scientific advisory capabilities, and it should recruit scientists associated with these national capabilities for UN scientific advisory functions. The United Nations will be better able to use scientific advice when all nations have the capability to participate fully in its scientific advisory processes.

Established Independent Science Organizations Offer an Excellent Means of Obtaining Advice.

The United Nations has a history of working with outside bodies to obtain information and advice on many issues chief executives and deliberative assembly bodies must address, including questions of environmental and social policy. Recognizing that implementation of Recommendations 1 and 2 may take some time to implement fully, chief executives and assemblies should, in the meantime, make arrangements with outside bodies to provide science advice. The International Council for Science (ICSU) and the InterAcademy Council, among others, are organizations that could undertake studies in response to specific questions posed by chief executives or assemblies. However, to ensure that the advice has the greatest degree of scientific credibility, the outside bodies should be required to operate on the basis of the elements discussed in Chapter 2 of this report.

Recommendation 4. To complement their internal scientific advisory processes, chief executives and deliberative assemblies, separately or in cooperation, should commission science policy advice from established independent organizations that follow procedures similar to those described here.

Scientific Assessment Mechanisms That Have Scientific Credibility and Transparency Are Especially Important to the Future Effectiveness of the UN System.

The Intergovernmental Panel on Climate Change (IPCC) is one such model of placing scientific assessments into the deliberative processes of a political assembly. These are surveys and analyses of the status of one or more important global problems, often with recommendations for international cooperative action. Part of the value of the IPCC rests on the credibility of its process. Its periodic reports have provided established information on trends in climate change to parties adhering to the UNFCCC, together with recommendations for actions that could be adopted in response. Other deliberative bodies could benefit from similar scientific and policy assessments that were undertaken on a periodic basis.

Recommendation 5. Assemblies and other deliberative bodies should make greater use of scientific assessment mechanisms, such as the IPCC, that have the transparency and credibility of a scientific process. Scientific assessment mechanisms provide a good model to be considered for other nonscientific, deliberative, and advisory processes.

In making these five recommendations, we recognize that each UN organization must adapt them to the circumstances of its own charters and operations. We also recognize that they cannot be implemented overnight and that they represent goals to be achieved by UN organizations over the first years of the 21st century.

The 20th century will be remembered for its great scientific and technological advances and their application to human welfare, especially in health, agriculture, and reduction of oppressive labor.[1] Tragically, the use of some of these technological powers for destructive purposes has become a clear threat to human civilization. The challenge at the dawn of the 21st century is to harness the ever-growing power of science and technology for improving the human condition and the well being of the Earth's life support systems for current and future generations. To fulfill this sustainability promise will require the mobilization of the scientific community of the world to contribute to this effort. One means for doing this is to make greater use of science and technology advice to the UN system in order to contribute to improved global governance. By strengthening its science advisory processes, the UN system will ensure its effective leadership in reaching goals of sustainability during the century ahead.

---

[1] A committee of the U.S. National Academy of Engineering selected these developments as the most important engineering accomplishments of the 20th century: electrification; automobile; airplane; water supply and distribution; electronics; radio and television; agricultural mechanization; computers; telephone; air conditioning and refrigeration; highways; spacecraft; Internet; imaging; household appliances; health technologies; petroleum and petrochemical technologies; laser and fiber optics; nuclear technologies; and high-performance materials. www.greatachievements.org.

REFERENCES

Agrawala, S. (1998a). "Context and early origins of the intergovernmental panel on climate change." *Climatic Change,* 39(4):605-620.

Agrawala, S. (1998b). "Structural and process history of the intergovernmental panel on climate change." *Climatic Change*, 39(4):621-642.

Andresen, S., Skodvin, T., Underdal, A. and Wettstad, J. eds. (2000), *Science and Politics in International Environmental Regimes*, Manchester University Press, Manchester, UK.

Auer, M. (1998). "Colleagues or Combatants? Experts as Environmental Diplomats," *International Negotiation*, 3(2):267-287.

Benedick, R.E. (1991). *Ozone diplomacy*. Cambridge: Harvard University Press.

Biermann, F. (2000). *Science as Power in International Environmental Negotiations: Global Environmental Assessments Between North and South*. ENRP Discussion Paper 2000-17. Belfer Center for Science and International Affairs, Kennedy School of Government, Harvard University, Cambridge, MA, USA.Bolin, B. (1994). "Science and policy making." *Ambio* 23(1).

Biermann, F. (2002). "Institutions for Scientific Advice: Global Environmental Assessments and Their Influence in Developing Countries," *Global Governance*, 8(2):195-219.

Binswanger, H. (2001). "Technological Progress and Sustainable Development: What about the Rebound Effect?" *Ecological Economics*, Vol. 36, pp. 119-132

Bolin, B. (1994). "Sciebce and Policy Making." *Ambio,* Vol, 23, No. 1.

Boyer, S. (2000). *Conference Diplomacy and UN Rules of Procedure*. Paper Presented at the Miami Workshop on Negotiation Skills for Climate Issues, July 24, Miami, FL.

Braithway, J. and Drahos, P. (2000). *Global Business Regulation*. Cambridge, UK. Cambridge University Press.

Carnegie Commission on Science, Technology, and Government. (1992). *International Environmental Research and Assessment: Proposals for Better Organization and Decision Making*. New York: Carnegie Corporation.

Chasek, P.S. (2001a). *Earth Negotiations: Analyzing Thirty Years of Environmental Diplomacy*. Tokyo, Japan and New York: United Nations University Press.

Chasek, P.S. (2001b). "Scientific Uncertainty in Environmental Negotiations" in H.-W. Jeong, ed. *Global Environmental Policies*. London, UK: Palgrave.

Chayes, A. and Chayes, A.H. (1995). *The New Sovereignty: Compliance with International Regulatory Agreements*. Cambridge, MA: Harvard University Press.

Conca, K., Alberty, M. and Dabelko, G. (1995). *Green Planet Blues: Environmental Politics from Stockholm to Rio*. Boulder, CO: Westview Press.

Corell, E. (1999a). "The Negotiable Desert: Expert Knowledge in the Negotiations of the Convention to Combat Desertification. PhD Dissertation." Department of Water and Environmental Studies, Linkoping University, Linkoping, Sweden.

Correll, E. (1999b). "Non-State Actor Influence in the Negotiations of the Convention to Combat Desertification," *International Negotiation*, Vol. 4, pp. 197-223.

Curlie, M. and Andresen, S. (2002). "International Trade in Endangered Species: The CITES Regime," in Miles, Edward L., Underdal, A., Andresen, S., Wettestad, J., Skjærseth, J., and Carlin, E. (2002). Eds. *Environmental Regime Effectiveness: Confronting Theory with Evidence*. Cambridge, MA and London, UK: MIT Press, pp. 357-379.

Davies, A. (1990). *Forty Years of Progress and Achievement—A Historic Review of WMO*. World Meteorological Organization, Geneva.

Dowdeswell, E. (2001). "A Global Dialogue: Avoiding a Genomics Divide." Paper Prepared for the Joint Centre for Bioethics, University of Toronto, Toronto, Canada.

Ezrahi, Y. (1990). *The Decent of Icarus: Science and the Transformation of Contemporary Democracy*. Cambridge, MA: Harvard University Press.

Fleagle, R.G. (1994). *Global environmental change: Interactions of science, policy and politics in the United States*. Westport, CT: Praeger.

Fritz, J. (2000). *Second Report on the International Advisory Process on the Environment and Sustainable Development*. UN System-Wide Earthwatch Coordination, United Nations Environment Programme, Geneva.

GEF (2002). *Adding Value to STAP Contributions to GEF Operations*. GEF Council Meeting, May 15-17, 2002. Global Environment Facility, Washington, DC.

Golden, W.T. (1991). *Worldwide Science and Technology Advice to the Highest Levels of Governments*. New York: Pergamon Press.

Gosovic, B. (1992). *The Quest for World Environmental Cooperation: The Case of the UN Global Environment Monitoring System*. London, UK and New York: Routledge.

Guagitsch, H. (2002). "Scientific Aspects of the Biosafety Debate," in Bail, C., Falkner, R., and Marquard, H., eds. (2002). *The Cartagena Protocol on Biosafety: Reconciling Trade and Biotechnology with Environment and Development?* London, UK: Earthscan Publishing and Royal Institute of International Affairs, pp. 83-91.

Gupta, A. (2000). "Governing Trade in Genetically Modified Organisms: The Cartagena Protocol on Biosafety," *Environment*, 42(4):22-33.

Haas, P.M. (1992). "Introduction: Epistemic Communities and International Policy Coordination," *International Organization*, 46(1):1-37.

Haas, P. M., Keohane, R. and Levy, M. (1993). *Institutions for the Earth: Sources of Effective International Environmental Protection*. Cambridge, MA and London, UK: MIT Press.

Heywood, V. and Watson, R. (1995), *Global Biodiversity Outlook*, Cambridge, UK: Cambridge University Press,

Hisschemöller, M. Hoppe, R., Dunn, W. and Ravetz, J. (2001). *Knowledge, Power, and Participation in Environmental Policy Analysis*. New Brunswick, NJ and London, UK: Transaction Publishers. (Policy Studies Review Annual, Vol.12)

Hoppe, R. (1999). "Policy Analysis, Science and Politics: From 'Speaking Truth to Power' to 'Making Sense Together'," *Science and Public Policy*, 26(3):201-210.

Intergovernmental Panel on Climate Change (2001), *Third Assessment Report*, World Meteorological Organization, Geneva.

Jasanoff, S. and Wynne, B. (1998). "Science and Decisionmaking," in Rayner, S. and Malone, E.L. eds. *Human Choices and Climate Change*. Columbus, OH: Battelle Press.

Juma, C. (2000)."The UN's Role in the New Diplomacy," *Issues in Science and Technology*, Vol. XVII, No. 1, Fall 2000, pp. 37-38.

Juma, C. (2002a). Science and Technology Diplomacy: Concepts and Elements of Work Program. Report Prepared for the United Nations Conference on Trade and Development, Geneva.

Juma, C. (2002b). "The Global Sustainability Challenge: From Agreement to Action," *International Journal of Environmental Issues*, 2(1/2):1-14.

Juma, C. and Henne, G. (1997). "Science and Technology in the Convention on Biological Diversity," in Raven, P.H. and Williams, T., eds. *Nature and Human Society: The Quest for a Sustainable World*. Washington, DC: National Academy Press, pp. 387-397.

Kaufman, J. (1996). *Conference Diplomacy: An Introductory Analysis*. London, UK, Macmillan.

Keck, M.E. and Sikkink, K. (1999). "Transnational advocacy networks in international and regional politics." *International Social Science Journal*, 51(1):89-101.

Knight, A. (2000), *Changing United Nations: Multilateral Evolution and the Quest for Global Governance.* London, UK: Macmillan.

Litfin, K.T. (1994). *Ozone Discourses: Science and Politics in Global Environmental Cooperation.* New York: Columbia University Press.

Litfin, K.T., ed. (1998a). *The Greening of Sovereignty in World Politics.* Cambridge, Massachusetts and London, UK: MIT Press.

Litfin, K.T. (1998b). "Satellites and Sovereign Knowledge: Remote Sensing of the Global Environment," in Litfin, K.T., ed., *The Greening of Sovereignty in World Politics.* Cambridge, MA and London, UK: MIT Press, pp. 193-221.

Marchant, G. (2002), "Biotechnology and the Precautionary Principle: Right Question, Wrong Answer," *International Journal of Biotechnology,* 4(1):34-45.

Marton-Lefevre, J. (1994), "The Role of the Scientific Community in the Preparation of and Follow-up to UNCED," in Spector, B., Sjostedt, G. and Zartman, I.W. eds. *Negotiating International Regimes: Lessons from the United Nations Conference on Environment and Development.* London, UK: Graham and Trotman.

Miles, E.L., Underdal, A., Andresen, S., Wettestad, J., Skjærseth, J., and Carlin, E. eds. (2002). *Environmental Regime Effectiveness: Confronting Theory with Evidence.* Cambridge, MA and London, UK: MIT Press.

Mitchell, R.B. (1998). "Forms of Discourse/Norms of Sovereignty: Interests, Science, and Morality in the Regulation of Whaling," in Litfin, K., ed., *The Greening of Sovereignty in World Politics.* Cambridge, MA: MIT Press, pp. 141-171.

National Research Council (1994). *Joint Statement, Population Summit of the World's Scientific Academies.* Washington, DC: National Academy Press, 1994, and, *Population -- the complex reality.* London: The Royal Society.

National Research Council (1995). *Standards, Conformity Assessment, and Trade: Into the 21st Century.* Washington, DC: National Academy Press.

National Research Council (1999a). *Our Common Journey: A Transition Toward Sustainability.* Washington, DC: National Academy Press.

National Research Council (1999b), *The Pervasive Role of Science, Technology, and Health in Foreign Policy: Imperatives for the Department of State.* Washington, DC: National Academy Press.

National Research Council (2000). *Incorporating Science, Economics, and Sociology in Developing Sanitary and Phytosanitary Standards in International Trade.* Proceedings of a Conference. Washington, DC: National Academy Press.

National Research Council (2002). *Down to Earth: Geographical Information for Sustainable Development in Africa."* Washington, DC: National Academy Press.

Parson, E.A. (Forthcoming). *Protecting the Ozone Layer: Science and Strategy.* Oxford, UK: Oxford University Press.

Posey, D. ed., (1999). *Cultural and Spiritual Values of Biodiversity.* London, UK: Intermediate Technology Publishers.

Ramsar Bureau (2002). Modus Operandi *of the Scientific and Technical Review Panel.* Draft Resolution. 8th Meeting of the Conference of the Contracting Parties to the Convention on Wetlands (Ramsar, Iran, 1971), Ramsar Bureau, Grand, Switzerland.

Raustiala, K. (1997). "States, NGOs and International Environmental Institutions." *International Studies Quarterly*, 41, 719-740.

Ravetz, J. (2001). "Viewpoint: Science Advice in the Knowledge Economy," *Science and Public Policy*, 28 (5):389-393.

Sagasti, F. (1984). "Reflections on the United Nations Conference on Science and Technology for Development," in Morehouse, W. Ed., *Third World Panacea or Global Boondoggle: The UN Conference on Science and Technology for Development Revisited,* Lund, Research Policy Institute, Discussion Paper No. 159, University of Lund, Lund, Sweden.

Sagasti, F. 1999. *Science and Technology in the United Nations System: An Overview*, Paper Prepared for the United Nations Development Programme, New York.

Saner, R. (2000). *Expert Negotiators.* Kluwer Academic Publisher, The Hague.

Schechter, M. ed. (1991). *United Nations-Sponsored World Conferences: Focus on Impact and Follow-Up,* Tokyo: United Nations University Press.

Sjostedt, G. (1994). "Issue Clarification and the Role of Consensual Knowledge in the UNCED System," in Spector, B., Sjostedt, G. and Zartman, I.W. eds. *Negotiating International Regimes: Lessons from the United Nations Conference on Environment and Development.* London, UK: Graham and Trotman.

Skodvin, T. (2000). *Structure and Agent in the Scientific Diplomacy of Climate Change: An Empirical Case Study of Science-Policy Interaction in the Intergovernmental Panel on Climate Change.* Dordrecht, The Netherlands: Kluwer Academic Publishers.

Skodvin, T. and Underdal, A. (2000). "Exploring the Dynamics of the Science-Politics Interactions," in Andresen, S., Skodvin, T., Underdal, A. and Wettstad, J. eds., *Science and Politics in International Environmental Regimes.* Manchester, UK: Manchester University Press, pp. 22-34.

Smith, B.L. (1992). *The Advisors: Scientists in the Policy Process*, Washington, DC: Brookings Institution.

The Social Learning Group. (2001a). *Learning to Manage Global Environmental Risks: A Comparative History of Social Responses to Climate Change, Ozone Depletion, and Acid Rain.* Vol. 1. Cambridge, MA and London, UK: MIT Press.

The Social Learning Group. (2001b). *Learning to Manage Global Environmental Risks: A Functional Analysis of Social Responses to Climate Change, Ozone Depletion, and Acid Rain* Vol. 2. Cambridge, MA and London, UK: MIT Press.

Susskind, L.E. (1994). *Environmental Diplomacy: Negotiating More Effective Global Agreements.* New York, New York and Oxford, UK: Oxford University Press.

Tolba, M. with Rummel-Bulska, I. (1998). *Global Environmental Diplomacy: Negotiating Environmental Agreements, 1973-1992.* Cambridge, MA: MIT Press.

Underdal, A. (2002). "Conclusion," in Miles, Edward L., Underdal, A., Andresen, S., Wettestad, J., Skjærseth, J., and Carlin, E. Eds. *Environmental Regime Effectiveness: Confronting Theory with Evidence.* Cambridge, MA and London, UK: MIT Press, pp. 433-465.

UNEP (1997), Modus Operandi *of the Subsidiary Body on Scientific, Technical and Technological Advice.* Secretariat of the Convention on Biological Diversity, United Nations Environment Programme, Nairobi.

UNEP (1999). *Synthesis of the Reports of the Scientific, Environmental Effects and Technology and Economic Assessment Panels of the Montreal Protocol. A Decade of Assessments for Decision Makers Regarding the Protection of the Ozone Layer: 1988-1999.* United Nations Environment Programme, Nairobi.

UNEP (2002). *Global Environment Outlook* 3, Earthscan, London, UK.

UNIDO (2002). *World Industrial Development report, 2000-2002.* United Nations Industrial Development Organization, Vienna

Victor, D.G., Raustiala, K. and Skolnikoff, E., eds. (1998). *The Implementation and Effectiveness of International Environmental Commitments: Theory and Practice.* Laxenburg, Austria: International Institute for Applied Systems Analysis and Cambridge, MA and London, UK: MIT Press.

Wagner, L. (1999). "Negotiations in the UN Commission on Sustainable Development: Coalitions, Processes, and Outcomes," *International Negotiation*, 4(3):107-131.

Wilkowski, J.M. (1982). *Conference Diplomacy II, A Case Study: The UN Conference on Science and Technology for Development, Vienna, 1979.* Washington, DC: Institute for the Study of Diplomacy, Georgetown University.

Willetts, P. (1996). "Consultative Status for NGOs at the United Nations," in Willetts, P. ed. *The Conscience of the World: The Influence of Non Governmental Organizations in the UN System*, London, UK: Hurst and Co.

World Commission on Culture and Development (1995), *Our Creative Diversity.* UNESCO, Paris.

World Commission on Dams (2002). *Dams and Development: A New Framework for Decision making.* Earthscan, London, UK.

World Commission on Environment and Development (1987). *Our Common Future.* Oxford, UK: Oxford University Press,

Young, O.R. (1999). *The Effectiveness of International Environmental Regimes: Causal Connections and Behavioral Mechanisms.* Cambridge, MA and London, UK: MIT Press.

ACRONYMS

ACAST Advisory Committee on the Application of Science and Technology for Development

ACFM Advisory Committee on Fishery Management

ACFR Advisory Committee on Fisheries Research

ACSTD UN Advisory Committee on Science and Technology for Development

BAHC Biospheric Aspects of the Hydrological Cycle

BDP Bureau for Development Policy of UNDP

BIOLAC Programme for Biotechnology in Latin America and the Caribbean of United Nations University

CITES Convention on the International Trade in Endangered Species of Wild Fauna and Flora

CMS Convention on Migratory Species

COFI Committee on Fisheries

COP Conference of the Parties

COSPAR Committee on Space Research

CSD Commission on Sustainable Development

CSTD Committee on Science and Technology for Development

DESA United Nations Department of Economic and Social Affairs

EAP Energy and Atmosphere Program of UNDP

ECG Ecosystem Conservation Group of UNEP

ECOSOC United Nations Economic and Social Council

FAO Food and Agriculture Organization

GEEP Group of Experts on the Effects of Pollutants

GEF Global Environment Facility

GEMS Global Earth Monitoring System of UNEP

GEMSI Group of Experts on Methods, Standards and Intercalibration

GESAMP Joint Group of Experts on the Scientific Aspects of Marine Environment Protection

GESREM Group of Experts on Standards and Reference Materials

GIPME Global Investigation of Pollution in the Marine Environment
GLODIR Global Directory of Marine and Freshwater Professionals of IOC
GOOS Global Ocean Observing System of IOC
HDRO Human Development Report Office in the United Nations Development Programme
IAC InterAcademy Council
IAP InterAcademy Panel for International Issues
IAS Institute of Advance Study of the United Nations University
ICAO International Civil Aviation Organization
ICES International Council for the Exploration of the Sea
ICGEB International Center for Genetic Engineering and Biotechnology
ICSU International Council for Science
IEA International Energy Agency
IGBP International Geosphere-Biosphere Programme
IGY International Geophysical Year
IHD International Hydrological Decade
IHP International Hydrological Programme
IMO International Maritime Organization
IOC Intergovernmental Oceanographic Commission
IODE International Oceanographic Data and Information Exchange System
IPCC Inter-governmental Panel on Climate Change
ISO International Organization for Standardization
ITU International Telecommunications Union
IUCN World Conservation Union
MA Millennium Ecosystem Assessment
NCEAS National Center for Ecological Analysis and Synthesis
NEAFC North-East Atlantic Fisheries Commission
NGO Nongovernmental organization
OECD Organization for Economic Cooperation and Development
SBSTTA Subsidiary Body on Scientific, Technical and Technological Advice
SCAR Scientific Committee on Antarctic Research
SCOPE Scientific Committee on Problems of the Environment

SCOR Scientific Committee on Oceanographic Research

SCOWAR Scientific Committee on Water Research

SEED Sustainable Energy and Environment Division of UNDP

SEI Stockholm Environment Institute

STAP Scientific And Technical Advisory Panel of the Global Environment Facility

UN United Nations

UNCED United Nations Conference on Environment and Development

UNCSEAR United Nations Committee on the Effects of Atomic Radiation

UNCSTD UN Conference on Science and Technology for Development

UNCTAD United Nations Conference on Trade and Development

UNDP United Nations Development Programme

UNEP United Nations Environment Programme

UNESCO United Nations Educational, Scientific, and Educational Organization

UNFCCC United Nations Framework Convention on Climate Change UNFSSTD UN Financing System for Science and Technology for Development

UNITAR United Nations Institute for Training and Research

UNU United Nations University

UNSCEAR United Nations Scientific Committee on the Effects of Atomic Radiation

WEA World Energy Assessment

WEC World Energy Council

WIPO World Intellectual Property Organization

WMO World Meteorological Organization

WRI World Resources Institute

APPENDIX I: THE UNITED NATIONS SYSTEM

NOTE: THESE EXCERPTS ARE IN THEIR ORIGINAL FORMAT AND WERE NOT EDITED.

SCIENCE ADVICE IN THE UNITED NATIONS SYSTEM

The UNITED NATIONS system

PRINCIPAL ORGANS OF THE UNITED NATIONS

- INTERNATIONAL COURT OF JUSTICE
- SECURITY COUNCIL
- GENERAL ASSEMBLY
- TRUSTEESHIP COUNCIL
- ECONOMIC AND SOCIAL COUNCIL
- SECRETARIAT

Security Council
- Military Staff Committee
- Standing Committee and ad hoc bodies
- International Criminal Tribunal for the Former Yugoslavia
- International Criminal Tribunal for Rwanda
- UN Monitoring, Verification and Inspection Commission (Iraq)
- United Nations Compensation Commission
- Peacekeeping Operations and Missions

General Assembly
- Main committees
- Other sessional committees
- Standing committees and ad hoc bodies
- Other subsidiary organs

PROGRAMMES AND FUNDS

UNCTAD United Nations Conference on Trade and Development
- **ITC** International Trade Centre (UNCTAD/WTO)

UNDCP United Nations Drug Control Programme

UNEP United Nations Environment Programme

UNHSP United Nations Human Settlements Programme (UN-Habitat)

UNDP United Nations Development Programme
- **UNIFEM** United Nations Development Fund for Women
- **UNV** United Nations Volunteers

UNFPA United Nations Population Fund

UNHCR Office of the United Nations High Commissioner for Refugees

UNICEF United Nations Children's Fund

WFP World Food Programme

UNRWA** United Nations Relief and Works Agency for Palestine Refugees in the Near East

OTHER UN ENTITIES

OHCHR Office of the United Nations High Commissioner for Human Rights

UNOPS United Nations Office for Project Services

UNU United Nations University

UNSSC United Nations System Staff College

RESEARCH AND TRAINING INSTITUTES

INSTRAW International Research and Training Institute for the Advancement of Women

UNICRI United Nations Interregional Crime and Justice Research Institute

UNITAR United Nations Institute for Training and Research

UNRISD United Nations Research Institute for Social Development

UNIDIR** United Nations Institute for Disarmament Research

FUNCTIONAL COMMISSIONS
- Commission for Social Development
- Commission on Human Rights
- Commission on Narcotic Drugs
- Commission on Crime Prevention and Criminal Justice
- Commission on Science and Technology for Development
- Commission on Sustainable Development
- Commission on the Status of Women
- Commission on Population and Development
- Statistical Commission

REGIONAL COMMISSIONS
- Economic Commission for Africa (ECA)
- Economic Commission for Europe (ECE)
- Economic Commission for Latin America and the Caribbean (ECLAC)
- Economic and Social Commission for Asia and the Pacific (ESCAP)
- Economic and Social Commission for Western Asia (ESCWA)
- United Nations Forum on Forests
- Sessional and Standing Committees
- Expert, ad hoc and related bodies

RELATED ORGANIZATIONS

IAEA International Atomic Energy Agency

WTO (trade) World Trade Organization

WTO (tourism) World Tourism Organization

CTBTO Preparatory Commission Preparatory Commission for the Comprehensive Nuclear-Test-Ban-Treaty Organization

OPCW Organization for the Prohibition of Chemical Weapons

SPECIALIZED AGENCIES*

ILO International Labour Organization

FAO Food and Agriculture Organization of the United Nations

UNESCO United Nations Educational, Scientific and Cultural Organization

WHO World Health Organization

WORLD BANK GROUP
- **IBRD** International Bank for Reconstruction and Development
- **IDA** International Development Association
- **IFC** International Finance Corporation
- **MIGA** Multilateral Investment Guarantee Agency
- **ICSID** International Centre for Settlement of Investment Disputes

IMF International Monetary Fund

ICAO International Civil Aviation Organization

IMO International Maritime Organization

ITU International Telecommunication Union

UPU Universal Postal Union

WMO World Meteorological Organization

WIPO World Intellectual Property Organization

IFAD International Fund for Agricultural Development

UNIDO United Nations Industrial Development Organization

Secretariat

- **OSG** Office of the Secretary-General
- **OIOS** Office of Internal Oversight Services
- **OLA** Office of Legal Affairs
- **DPA** Department of Political Affairs
- **DDA** Department for Disarmament Affairs
- **DPKO** Department of Peacekeeping Operations
- **OCHA** Office for the Coordination of Humanitarian Affairs
- **DESA** Department of Economic and Social Affairs
- **DGAACS** Department of General Assembly Affairs and Conference Services
- **DPI** Department of Public Information
- **DM** Department of Management
- **OIP** Office of the Iraq Programme
- **UNSECOORD** Office of the United Nations Security Coordinator
- **ODCCP** Office for Drug Control and Crime Prevention
- **UNOG** UN Office at Geneva
- **UNOV** UN Office at Vienna
- **UNON** UN Office at Nairobi

*Autonomous organizations working with the United Nations and each other through the coordinating machinery of the Economic and Social Council.
**Report only to the General Assembly.

Published by the United Nations Department of Public Information
DPI/2079/Add.1 - January 2002

APPENDIX II: Procedures for the Preparation, Review, Acceptance, Adoption, Approval, and Publication of IPCC Reports

Excerpts

4. ASSESSMENT REPORTS, SYNTHESIS REPORTS, SPECIAL REPORTS AND METHODOLOGY GUIDELINES

4.1 Introduction to Review Process

The review process generally takes place in three stages: expert review of IPCC Reports, government/expert review of IPCC Reports, and government review of the Summaries for Policymakers and/or the Synthesis Report. Working Group Co-chairs should aim to avoid (or at least minimize) the overlap of government review periods for different IPCC Reports and with Sessions of the Conference of Parties of the United Nations Framework Convention of Climate Change and its subsidiary bodies.

Expert review should normally be eight weeks, but not less than six weeks, except to the extent decided by the Panel. Government and government/expert reviews should not be less than eight weeks, except to the extent decided by the Panel.

All written expert, and government review comments will be made available to reviewers on request during the review process and will be retained in an open archive in a location determined by the IPCC Secretariat on completion of the Report for a period of at least five years.

4.2 Reports Accepted by Working Groups

Reports presented for acceptance at Sessions of the Working Groups are the full scientific, technical and socio-economic Assessment Reports of the Working Groups, Special Reports and Methodology Guidelines, such as the IPCC Guidelines for National Greenhouse Gas Inventories or the IPCC Technical Guidelines for Assessing Climate Change Impacts and Adaptations.

The subject matter of these Reports shall conform to the terms of reference of the relevant Working Groups and to the work plan approved by the Panel.

Reports to be accepted by the Working Groups will undergo expert and government/expert reviews. The purpose of these reviews is to ensure that the Reports present a comprehensive, objective, and balanced view of the areas they cover. While the large volume and technical detail of this material places practical limitations upon the extent to which changes to these Reports will normally be made at Sessions of Working

Groups, "acceptance" signifies the view of the Working Group that this purpose has been achieved. The content of the authored chapters is the responsibility of the Lead Authors, subject to Working Group acceptance.

Changes (other than grammatical or minor editorial changes) made after acceptance by the Working Group shall be those necessary to ensure consistency with the Summary for Policymakers. These changes shall be identified by the Lead Authors in writing and made available to the Panel at the time it is asked to accept the Summary for Policymakers.

Reports accepted by Working Groups should be formally and prominently described on the front and other introductory covers as:

"A report accepted by Working Group X of the IPCC but not approved in detail."

It is essential that Working Group work programmes allow enough time in their schedules, according to procedures, for a full review by experts and governments and for the Working Group's acceptance. The Working Group Co-Chairs are responsible for implementing the work programme and ensuring that proper review of the material occurs in a timely manner.

To ensure proper preparation and review, the following steps should be undertaken:

1. Compilation of lists of Coordinating Lead Authors, Lead Authors, Contributing Authors, Expert Reviewers, Review Editors and Government Focal Points.

2. Selection of Lead Authors.

3. Preparation of draft Report.

4. Review.
    a. First review (by experts).
    b. Second review (by governments and experts).

5. Preparation of final draft Report.

6. Acceptance of Report at a Session of the Working Group(s).

4.2.1 Compilation of Lists of Coordinating Lead Authors, Lead Authors, Contributing Authors, Expert Reviewers, Review Editors and Government Focal Points

At the request of Working Group Co-Chairs through their respective Working Group Bureau and the IPCC Secretariat, governments, and participating organizations and the Working Group Bureaus should identify appropriate experts for each area in the Report who can act as potential Coordinating Lead Authors, Lead Authors, Contributing Authors, expert reviewers or Review Editors. To facilitate the identification of experts and later review by governments, governments should also designate their respective

Focal Points. IPCC Bureau Members should contribute where necessary to identifying appropriate Coordinating Lead Authors, Lead Authors, Contributing Authors, expert reviewers, and Review Editors in cooperation with the Government Focal Points within their region to ensure an appropriate representation of experts from developing and developed countries and countries with economies in transition. These should be assembled into lists available to all IPCC Members and maintained by the IPCC Secretariat. The tasks and responsibilities of Coordinating Lead Authors, Lead Authors, Contributing Authors, expert reviewers, Review Editors and government Focal Points are outlined in Annex 1.

4.2.2 Selection of Lead Authors

Coordinating Lead Authors and Lead Authors are selected by the relevant Working Group Bureau, under general guidance and review provided by the Session of the Working Group, from those experts cited in the lists provided by governments and participating organizations, and other experts as appropriate, known through their publications and works. The composition of the group of Coordinating Lead Authors and Lead Authors for a section or chapter of a Report shall reflect the need to aim for a range of views, expertise and geographical representation (ensuring appropriate representation of experts from developing and developed countries and countries with economies in transition). There should be at least one and normally two or more from developing countries. The Coordinating Lead Authors and Lead Authors selected by the Working Group Bureau may enlist other experts as Contributing Authors to assist with the work.

At the earliest opportunity, the IPCC Secretariat should inform all governments and participating organisations who the Coordinating Lead Authors and Lead Authors are for different chapters and indicate the general content area that the person will contribute to the chapter.

4.2.3 Preparation of Draft Report

Preparation of the first draft of a Report should be undertaken by Coordinating Lead Authors and Lead Authors. Experts who wish to contribute material for consideration in the first draft should submit it directly to the Lead Authors. Contributions should be supported as far as possible with references from the peer-reviewed and internationally available literature, and with copies of any unpublished material cited. Clear indications of how to access the latter should be included in the contributions. For material available in electronic format only, a hard copy should be archived and the location where such material may be accessed should be cited.

Lead Authors will work on the basis of these contributions, the peer-reviewed and internationally-available literature, including manuscripts that can be made available for IPCC review and selected non-peer review literature according to Annex 2 and IPCC Supporting Material (see section 6). Material which is not published but which is available to experts and reviewers may be included provided that its inclusion is fully justified in the context of the IPCC assessment process (see Annex 2).

In preparing the first draft, and at subsequent stages of revision after review, Lead Authors should clearly identify disparate views for which there is significant scientific or technical support, together with the relevant arguments.

Technical summaries provided by the Working Groups will be prepared under the leadership of the Working Group Bureaus.

4.2.4 Review

Three principles governing the review should be borne in mind. First, the best possible scientific and technical advice should be included so that the IPCC Reports represent the latest scientific, technical and socio-economic findings and are as comprehensive as possible.

Secondly, a wide circulation process, ensuring representation of independent experts (i.e. experts not involved in the preparation of that particular chapter) from developing and developed countries and countries with economies in transition should aim to involve as many experts as possible in the IPCC process. Thirdly, the review process should be objective, open and transparent.

To help ensure that Reports provide a balanced and complete assessment of current information, the Bureau of each Working Group should normally select two Review Editors per chapter (including the executive summaries) and per technical summary of each Report.

Review Editors should normally consist of a member of the Working Group Bureau and an independent expert based on the lists provided by governments and participating organizations. Review Editors should not be involved in the preparation or review of material for which they are an editor. In selecting Review Editors, the Bureaus should select from developed and developing countries and from countries with economies in transition, and should aim for a balanced representation of scientific, technical, and socio-economic views.

4.2.4.1 First Review (by Experts)

First draft Reports should be circulated by Working Group Co-Chairs for review by experts selected by the Working Group Bureaus and, in addition, those on the lists provided by governments and participating organizations, noting the need to aim for a range of views, expertise, and geographical representation. The review circulation should include:

• Experts who have significant expertise and/or publications in particular areas covered by the Report.

- Experts nominated by governments as Coordinating Lead Authors, Lead Authors, contributing authors or expert reviewers as included in lists maintained by the IPCC Secretariat.

- Expert reviewers nominated by appropriate organizations.

The first draft Reports should be sent to Government Focal Points, for information, along with a list of those to whom the Report has been sent for review in that country.

The Working Group Co-Chairs should make available to reviewers on request during the review process specific material referenced in the document being reviewed, which is not available in the international published literature.

Expert reviewers should provide the comments to the appropriate Lead Authors through the relevant Working Group Co-Chairs with a copy, if required, to their Government Focal Point.

Coordinating Lead Authors, in consultation with the Review Editors and in coordination with the respective Working Group Co-Chairs and the IPCC Secretariat, are encouraged to supplement the draft revision process by organizing a wider meeting with principal Contributing Authors and expert reviewers, if time and funding permit, in order to pay special attention to particular points of assessment or areas of major differences.

4.2.4.2 Second Review (by Governments and Experts)

A revised draft should be distributed by the appropriate Working Groups or through the IPCC Secretariat to governments through the designated Government Focal Points, and to all the coordinating lead authors, lead authors and contributing authors and expert reviewers.

Governments should send one integrated set of comments for each Report to the appropriate Working Group through their Government Focal Points.

Non-government reviewers should send their further comments to the appropriate Working Group Co-Chairs with a copy to their appropriate Government Focal Point.

4.2.5 Preparation of Final Draft Report

Preparation of a final draft Report taking into account government and expert comments for submission to a Session of a Working Group for acceptance should be undertaken by Coordinating Lead Authors and Lead Authors in consultation with the Review Editors. If necessary, and timing and funding permitting, a wider meeting with principal Contributing Authors and expert and government reviewers is encouraged in order to pay special attention to particular points of assessment or areas of major differences. It is important that Reports describe different (possibly controversial) scientific, technical, and socio-economic views on a subject, particularly if they are relevant to the policy debate.

The final draft should credit all Coordinating Lead Authors, Lead Authors, Contributing Authors, reviewers and Review Editors by name and affiliation (at the end of the Report).

4.3 Approval and Acceptance of Summaries for Policymakers

Summary sections of Reports approved by the Working Groups and accepted by the Panel will principally be the Summaries for Policymakers, prepared by the respective Working Groups of their full scientific, technical and socio-economic assessments, and Summaries for Policymakers of Special Reports prepared by the Working Groups. The Summaries for Policy Makers should be subject to simultaneous review by both experts and governments and to a final line by line approval by a Session of the Working Group. Responsibility for preparing first drafts and revised drafts of Summaries for Policymakers, lies with the respective Working Group Co-Chairs. The Summaries for Policymakers should be prepared concurrently with the preparation of the main Reports.

Approval of the Summary for Policymakers at the Session of the Working Group, signifies that it is consistent with the factual material contained in the full scientific, technical and socio-economic assessment or Special Report accepted by the Working Group. Coordinating lead authors may be asked to provide technical assistance in ensuring that consistency has been achieved. These Summaries for Policymakers should be formally and prominently described as:

"A Report of [Working Group X of] the Intergovernmental Panel on Climate Change."

For a Summary for Policymakers approved by a Working Group to be endorsed as an IPCC Report, it must be accepted at a Session of the Panel. Because the Working Group approval process is open to all governments, Working Group approval of a Summary for Policymakers means that the Panel cannot change it. However, it is necessary for the Panel to review the Report at a Session, note any substantial disagreements, (in accordance with Principle 10 of the Principles Governing IPCC Work) and formally accept it.

4.4 Reports Approved and/or Adopted by the Panel

Reports approved and/or adopted by the Panel will be the Synthesis Report of the Assessment Reports and other Reports as decided by the Panel whereby Section 4.3 applies *mutatis mutandis*.

4.4.1 The Synthesis Report

The Synthesis Report will synthesize and integrate materials contained within the Assessment Reports and Special Reports and should be written in a non-technical style suitable for policymakers and address a broad range of policy-relevant but policy-neutral questions approved by the Panel. The Synthesis Report is composed of two sections as follows: (a) a Summary for Policymakers and (b) a longer report. The IPCC Chair will lead a writing team whose composition is agreed by the Bureau, noting the need to aim for a range of views, expertise and geographical representation.

An approval and adoption procedure will allow Sessions of the Panel to approve the SPM line by line and to ensure that the SPM and the longer report of the Synthesis Report are consistent, and the Synthesis Report is consistent with the underlying Assessment Reports and Special Reports from which the information has been synthesised and integrated. This approach will take 5-7 working days of a Session of the Panel.

Step 1: The longer report (30-50 pages) and the SPM (5-10 pages) of the Synthesis Report are prepared by the writing team.

Step 2: The longer report and the SPM of the Synthesis Report undergo simultaneous expert/government review.

Step 3: The longer report and the SPM of the Synthesis Report are then revised by Lead Authors, with the assistance of the Review Editors.

Step 4: The revised drafts of the longer report and the SPM of the Synthesis Report are submitted to Governments and participating organizations eight weeks before the Session of the Panel.

Step 5: The longer report and the SPM of the Synthesis Report are both tabled for discussion in the Session of the Panel:

- The Session of the Panel will first provisionally approve the SPM line by line.

- The Session of the Panel will review and adopt the longer report of the Synthesis Report, section by section, i.e. roughly one page or less at a time.

The review and adoption process for the longer report of the Synthesis Report should be accomplished in the following manner:

- When changes in the longer report of the Synthesis Report are required either to conform it to the SPM or to ensure consistency with the underlying Assessment Reports, the Panel and authors will note where changes are required in the longer report of the Synthesis Report to ensure consistency in tone and content. The authors of the longer report of the Synthesis Report will then make changes in the longer report of the Synthesis Report. Those Bureau members who are not authors will act as Review Editors to ensure that these documents are consistent and follow the directions of the Session of the Panel

- The longer report of the Synthesis Report is then brought back to the Session of the Panel for the review and adoption of the revised sections, section by section. If inconsistencies are still identified by the Panel, the longer report of the Synthesis Report is further refined by the Authors with the Assistance of the Review Editors for review and adoption by the Panel. This process is conducted section by section, not line by line.

- The final text of the longer report of the Synthesis Report will be adopted and the SPM approved by the Session of the Panel. The Report consisting of the longer report and the SPM of the Synthesis Report is an IPCC. Report and should be formally and prominently described as: "A Report of the Intergovernmental Panel on Climate Change."

APPENDIX III: *Modus Operandi* of the Subsidiary Body on Scientific, Technical and Technological Advice of the Convention on Biological Diversity

I.   Functions

1.   The functions of the SBSTTA are those contained in Article 25 of the Convention. Accordingly, the SBSTTA will fulfil its mandate under the authority of, and in accordance with, guidance laid down by the Conference of the Parties, and upon its request.

2.   Pursuant to Article 25, paragraph 3, the functions, terms of reference, organization and operation of the SBSTTA may be further elaborated, for approval by the Conference of the Parties.

II.   Rules of procedure

3.   The rules of procedure for meetings of the Conference of the Parties to the Convention on Biological Diversity shall apply, in accordance with rule 26, paragraph 5, *mutatis mutandis* to the proceedings of the SBSTTA. Therefore, rule 18 on credentials will not apply.

4.   In accordance with rule 52, the official and working languages of the SBSTTA will be those of the United Nations Organization. The proceedings of the SBSTTA will be carried out in the working languages of the Conference of the Parties.

5.   In order to facilitate continuity in the work of SBSTTA and taking into account the technical and scientific character of the input of SBSTTA, the terms of office of members of the Bureau of SBSTTA will be two years. At each meeting of the SBSTTA one of the two regional representatives shall be elected in order to achieve staggered terms of office. The members of the Bureau of SBSTTA will take office at the end of the meeting at which they are elected.

6.   The Chairman of the SBSTTA, elected at an ordinary meeting of the Conference of the Parties, shall take office from the beginning of the next ordinary meeting of the SBSTTA and remain in office until the beginning of the following meeting of the SBSTTA.

III. Frequency and timing of the SBSTTA

7. The SBSTTA shall meet annually and sufficiently in advance of each regular meeting of the Conference of the Parties, for a duration to be determined by the Conference of the Parties which should not normally exceed five days. The number and length of the meetings and activities of the SBSTTA and its organs should be reflected in the budget adopted by the Conference of the Parties or other sources of extra-budgetary funding.

IV. Documentation

8. The documentation prepared for meetings will be distributed six weeks before the meeting in the working languages of the SBSTTA, will be concrete, focused draft technical reports and will include proposed conclusions and recommendations for consideration of the SBSTTA.

9. To facilitate the preparation of documentation, and in order to avoid duplication of efforts and ensure the use of available scientific, technical and technological competence available within international and regional organizations, including non-governmental organizations and scientific unions and societies, qualified in fields relating to conservation and sustainable use of biodiversity, the Executive Secretary may establish, in consultation with the Chairman and the other members of the Bureau of the SBSTTA, liaison groups, as appropriate. Such liaison groups will depend on the resources available.

V. Organization of work during the meetings

10. Each meeting of the SBSTTA will propose to the Conference of the Parties, in light of the programme of work for the Conference of the Parties and the SBSTTA, a particular theme as the focus of work for the following meeting of the SBSTTA.

11. Two open-ended sessional working groups of the SBSTTA could be established and operate simultaneously during meetings of the SBSTTA. They shall be established on the basis of well-defined terms of reference, and will be open to all Parties and observers. The financial implications of these arrangements should be reflected in the budget of the Convention.

VI. Ad hoc technical expert group meetings

12. A limited number of ad hoc technical expert groups on specific priority issues on the programme of work of the SBSTTA may be established, as required, for a limited duration. The establishment of such ad hoc technical expert groups would be guided by the following elements:

(a) The ad hoc technical expert groups should draw on the existing knowledge and competence available within, and liaise with, international, regional and national organizations, including non-governmental organizations and the scientific community in fields relevant to this Convention;

(b) The ad hoc technical expert groups shall be composed of no more than fifteen experts competent in the relevant field of expertise, with due regard to geographical representation and to the special conditions of least-developed countries and small island developing States;

(c) The SBSTTA will recommend the exact duration and specific terms of reference, when establishing such expert groups for the approval of the Conference of the Parties;

(d) Expert groups will be encouraged to use innovative means of communication and to minimize the need for face-to-face meetings;

(e) The ad hoc technical expert groups may also convene meetings parallel to the proceedings of the SBSTTA;

(f) All efforts will be made to provide adequate voluntary financial assistance for the participation of experts in the expert groups from developing countries and countries with economies in transition Parties; and

(g) The number of ad hoc technical expert groups active each year will be limited to a maximum of three and will depend on the amount of resources designated to the SBSTTA by the Conference of the Parties in its budget or on the availability of extra-budgetary resources.

VII. Contribution of non-governmental organizations

13. The scientific and technical contribution of non-governmental organizations to the fulfilment of the mandate of the SBSTTA will be strongly encouraged in accordance with the relevant provisions of the Convention and the rules of procedure for meetings of the Conference of the Parties.

VIII. Cooperation with other relevant bodies

14. The SBSTTA shall cooperate with other relevant international, regional and national organizations, under the guidance of the Convention of the Parties, thus building upon the vast experience and knowledge available.

15. In this context, the SBSTTA emphasizes the importance of research to further increase available knowledge and reduce uncertainties, and recommends that the Conference of the Parties consider this issue in relation to the financial resources required for the effective implementation of the Convention.

IX. Regional and subregional preparatory meetings

16. Regional and subregional meetings for the preparation of regular meetings of the SBSTTA may be organized as appropriate for specific items. The possibility of combining such meetings with other scientific regional meetings, in order to make maximum use of available resources, should be considered. The convening of such regional and subregional meetings will be subject to the availability of voluntary financial contributions.

17. The SBSTTA should, in the fulfilment of its mandate, draw upon the contributions of the existing regional and subregional intergovernmental organizations or initiatives.

X. Focal points

18. A list of focal points and focal persons to the SBSTTA shall be established and regularly updated by the Secretariat, on the basis of information provided by Parties and other relevant regional, subregional and intergovernmental organizations.

XI. Roster of experts

19. A roster of experts, in the relevant fields of the Convention, will be compiled by the Secretariat on the basis of input from all Parties and, as appropriate, from other countries and relevant bodies. The roster of experts will be regularly updated and will be made accessible through the clearing house mechanism.

20. The ad hoc technical expert groups and liaison groups referred to above as well as the Secretariat should make full use of such a roster of experts, inter alia, through scientific peer review processes.

APPENDIX IV: Draft Resolution on the *Modus Operandi* of the Scientific and Technical Review Panel (STRP) of the Ramsar Convention on Wetlands

Excerpts

1. RECALLING the establishment by Resolution 5.5 of the Scientific and Technical Review Panel (STRP), made up of members with appropriate scientific and technical knowledge, appointed by the Conference of the Contracting Parties (COP), but participating as individuals and not as representatives of their countries of origin;

...

THE CONFERENCE OF THE CONTRACTING PARTIES

7. REAFFIRMS the critical importance to the Convention of the work and advice of the Scientific and Technical Review Panel (STRP) in providing reliable guidance to the Conference of the Contracting Parties;

8. APPROVES the revised *modus operandi* for the STRP as annexed to this Resolution, and DECIDES that the provisions in the Annex supersede those in the previous Resolutions on the STRP dealing with the same issues;

9. RECOGNIZES the urgent need to ensure both that the Panel is provided with the necessary resources to undertake its work effectively and efficiently and that the Ramsar Bureau has sufficient capacity to support this work, and URGES Contracting Parties and others to afford the highest priority for securing continuity of such funding;

...

11. REVISES as follows the list of bodies and organizations invited to participate as observers in the meetings of the STRP during the 2003-2005 triennium, in addition to the International Organization Partners, and INVITES them to consider establishing close working cooperative arrangements with the STRP on matters of common interest:

    - the Subsidiary Body on Scientific, Technical and Technological Advice of the Convention on Biological Diversity (CBD)
    - the Scientific Council of the Convention on Migratory Species (CMS)
    - the Committee on Science and Technology of the Convention to Combat Desertification (UNCCD)
    - the Subsidiary Body on Scientific and Technical Advice of the Framework Convention on Climate Change (UNFCCC)

- the secretariats of the CBD, CMS, UNCCD, UNFCCC and the Millennium Ecosystem Assessment
- the Society of Wetland Scientists
- [the International Association of Limnology]
- [the Global Wetlands Economics Network]
- the International Mire Conservation Group
- the International Peat Society
- the Center for International Earth Science Information Network (CIESIN), Columbia University, USA
- the International Association for Impact Assessment (IAIA)
- The Nature Conservancy (TNC)
- Ducks Unlimited (Canada, Mexico, and USA)
- The World Resources Institute (WRI)
- The Institute for Inland Water Management and Wastewater Treatment (RIZA) (The Netherlands)

12. FURTHER EMPHASIZES the value of participation by STRP members in meetings of the COP and Standing Committee, and REQUESTS Contracting Parties, the Standing Committee, and the Ramsar Bureau all to do their utmost to secure any additional funding which might be necessary for this purpose;

...

14. REAFFIRMS that the membership of the STRP shall have the same regional structure as the Standing Committee, as established in Resolution VII.1; and that the same proportional system that the Standing Committee will apply for determining its composition; and that, in order to attain equitable representation on the subsidiary bodies of the Convention, members of the STRP ought, as far as possible, to come from Contracting Parties different from those Parties elected to the Standing Committee;

...

[16. ENDORSES the establishment of a post in the Bureau to assist the Deputy Secretary General in the day-to-day support of the work of the Panel and its Working Groups and the development of the STRP National Focal Points network.]

Modus operandi of the Convention's Scientific and Technical Review Panel (STRP)

Establishing STRP tasks and priorities

1. The Conference of the Contracting Parties (COP) shall have available to it a list of STRP assignments, derived by the Bureau from the draft Convention Work Plan for the next triennium and draft Resolutions submitted to the COP. The COP shall establish the priorities for STRP work in the coming triennium.

2. The Standing Committee shall adopt the definitive list of STRP assignments for the triennium on the basis of the Convention Work Plan and Resolutions adopted by the COP, and will provide additional guidance on priority tasks.

3. At its first meeting, the STRP shall agree on its work plan and identify which tasks it considers can be undertaken in the triennium with the available resources and which ones will require additional resources for their implementation.

4. The Bureau shall circulate immediately by e-mail the work plan agreed by the STRP at its first meeting for comments by the Standing Committee and the STRP National Focal Points, with a deadline for inputs of three weeks, so as to ensure the maximum possible time for the STRP to undertake its work.

Schedule and purpose of meetings, and process between meetings

5. The STRP will meet twice in plenary during a triennium. The first meeting shall take place no later than six months after the COP, and the second meeting approximately nine months prior to the next COP.

6. The first meeting of the STRP in each triennium shall:

    a) include an 'induction and briefing session' for all participants in order to ensure that they (particularly members appointed for the first time) are fully aware of their respective roles and responsibilities prior to making decisions on progressing the work requested oPanel. The briefing will emphasize the role of the Panel in relation to that of the COP, the Standing Committee (to which the STRP reports), and the Ramsar Bureau.

    b) establish the STRP work plan for the triennium, based on the tasks and priorities identified by the COP and Standing Committee, taking into account also issues arising from the Panel's role in reviewing strategically current tools and guidance available to Parties, and new and emerging issues for the Convention;

    c) establish an expert Working Group for each substantive task in the STRP work plan, identify the members of each Working Group, and agree the modus operandi for each Working Group to undertake its tasks;

    d) identify additional experts to be invited to contribute to the work of each Working Group, either in the drafting of materials or in reviewing such materials. In so doing, the STRP should give due consideration to the geographical and gender balance, and to the language abilities of the proposed experts; and

e) identify key additional strategic issues for consideration by the STRP during the triennium, and establish a Working Group to progress these for reporting to the next COP.

7. Thereafter, expert Working Groups shall develop and undertake their work largely through electronic communication, tele- and video-conferences, virtual forums and exchange networks. The Bureau shall assist in establishing such mechanisms as necessary.

8. Each expert Working Group shall, as resources permit, meet in a workshop approximately nine months after the 1st STRP meeting, in order to review draft materials, amend their parts of the work plan as necessary, and agree the steps to be taken for timely completion of their tasks.

9. Where the STRP determines that it requires the expertise of an invited expert to prepare draft materials for its review, resources permitting, the Bureau shall arrange as necessary for contracts to be let immediately after the period for comments on the STRP work plan is over. First draft materials prepared under these contracts will be available for review by the relevant STRP Working Group before any mid-term Working Group workshop.

10. Should the dates for the next COP be set less than 3 calendar years (36 months) after the previous COP, the STRP Working Groups shall review their workloads and agreed deliverables, and advise the Standing Committee of any proposed changes to the Panel's work plan.

11. The second meeting of the Panel shall:

    a) receive reports from each of its expert Working Groups, including final draft guidelines and other materials;

    b) review and approve finalization of these materials for consideration by the Standing Committee and COP;

    c) identify any further work on each topic that it may consider is still needed, and make recommendations on this to the Standing Committee and COP; and

    d) review the recommendations of the Working Group on key strategic issues for the Convention, and prepare these for consideration by the Standing Committee and COP.

12. The working language of the Panel shall be English. The ability of STRP members and invited experts to consult and use literature in other languages shall constitute an additional asset for their appointment.

The roles and responsibilities of the Panel and its members

13. The Terms of Reference of the STRP and its Members are to:

a) review the tasks and nature of the products requested of it by COP Resolutions and the Convention's Work Plan;

b) undertake strategic review of the current tools and guidance available to Parties and new and emerging issues for the Convention;

c) determine and agree a mechanism for the delivery of each of these tasks, including the establishment of expert Working Groups as appropriate, advise on which tasks it does not have the expertise or capacity to progress, and receive the advice of the Standing Committee for this work plan;

d) identify, for each task the Panel proposes to undertake, and with the advice of any Working Group on the topic, the best global expert(s) either from within or outside the Panel, to undertake drafting work, taking into account geographical and gender balance and language ability;

e) identify, for each product in the work plan, and with the advice of any Working Group on the topic, additional experts to undertake review by correspondence of draft materials, as necessary;

f) make expert review of the draft products in its work plan, taking into account the views expressed by additional experts in d) above, agree any amendments needed, and transmit these revised products for consideration by the Standing Committee;

g) ensure, with the assistance of the Ramsar Bureau, that the work of the STRP contributes to and benefits from the work undertaken by similar subsidiary bodies of other multilateral environmental agreements (MEAs).

14. In undertaking their work, members of the STRP should, as set out in the ToR for STRP National Focal Points (NFPs), establish and maintain contact with the National Focal Points in their (sub) region, with an agreed allocation of Contracting Parties to each regional member, in order to ensure that the views and expertise of NFPs is available to the Panel.

15. In undertaking their work, International Organization Partner (IOP) members of the STRP should ensure that their networks, including their expert Specialist Groups, are consulted on the work of the Panel and that their views and expertise is available to the Panel.

16. STRP members should, as resources permit, participate in meetings of the COP and Standing Committee.

The role of STRP expert Working Groups and their Leads

17. Terms of Reference for expert Working Groups established by the STRP are:

    Under the guidance of the Working Group Lead, to:

    a) prepare a work plan for the Working Group tasks as identified by COP Resolutions, including scoping the structure and contents of any guidelines and reports and proposing a mechanism and timeframe for their delivery;

    b) review draft materials prepared under this work plan, and advise on any necessary revisions, amendments or further work; and

    c) advise the Panel when the Working Group's scientific and technical work on the guidelines and reports is complete, so that the materials can be recommended by the Panel to the Standing Committee for consideration.

18. The role and responsibilities of a Working Group Lead are to oversee and guide the work of the expert Working Group so as to ensure timely review and delivery of its products, including through electronic networking and chairing of any Working Group workshop. In undertaking this role the Working Group Lead will work closely with the Chair or Vice-Chair so as to keep the Chair or Vice-Chair advised on progress.

19. Appointment of Leads of Working Groups will be made by the Chair of the STRP with the assistance of the Bureau at the first meeting of the STRP in a triennium. A Working Group Lead need not necessarily be an STRP member, but could also come from an IOP or other observer organization or from among the invited experts.

20. A Working Group Lead should have proven international expertise in the theme of the Working Group and, ideally, previous experience of the modus operandi of the Convention and its bodies and the nature of scientific and technical materials required by the Convention.

21. Where a Working Group theme continues in the STRP work plan for more than one triennium, its Lead may, as appropriate, be appointed for a further term.

22. Working Group Leads should be prepared to represent the Panel in contributing to the work of equivalent expert working groups established by other MEAs on similar topics. Working Group Leads should recognise and confirm the acceptance of such potential time commitments at the time of their appointment.

...

The role of observer organizations

30. The primary role of observer organizations is to bring technical and scientific review capacity on their topics of expertise to the review work of the Panel. However, given the lead technical prowess of such organizations, it may be appropriate that a member or members of their network take the lead in the role of an 'invited expert' to undertake drafting work for the Panel.

31. STRP observer status shall be a consistent mechanism for engaging the involvement of all scientific and technical organizations with which the Convention develops formal collaborative agreements.

32. The Panel may request that representatives of other relevant scientific and technical organizations be invited as observers to STRP, as it deems necessary, in order to increase the capacity of the Panel in specific subject areas on which it is requested to work.

33. Each observer organization shall identify to the Chair of the STRP and the Bureau a named representative who will participate in the meetings and work of the Panel. An observer organization should be prepared to participate in all Panel meetings during a triennium and should send the same representative to these meetings, if possible.

…

APPENDIX V: Rules of Procedure of the InterAcademy Council

Excerpts

...

Article 3. PROCEDURES FOR THE CONDUCT ON AN IAC STUDY

1) Proposals for IAC studies may originate from a requesting agency (e.g., an international organization, national government or group of governments) or from the IAC Board itself.

2) Whenever a request is received for the IAC to conduct a study on a specific topic, the IAC Board co-Chairs shall request the IAC Executive Director to prepare a study prospectus -- in consultation with requesting agency and designated IAP Academies where appropriate -- including a budget for the completion of the proposed study. Requesting agencies shall also be asked to suggest a balance of national sources for the advice it needs. A report review procedure shall be incorporated into the study design, depending upon considerations of the study scope and timeframe. Estimates of cost of report publication shall be included in the proposal budget.

3) The study prospectus shall be presented to the IAC Board for review and decision, with a recommendation from the IAC Board co-Chairs. By two-thirds majority, the IAC Board shall approve [or disapprove] the study prospectus based on considerations of importance and timeliness of the question, background, likely impact, engagement of likely audiences, the range of competencies that must be represented on the study panel, dissemination mechanisms, and funding sources.

4) If the prospectus is approved by the IAC Board, a formal proposal shall be prepared by the IAC Executive Director and submitted to the requesting agency or other organizations that shall assume responsibility for providing financial resources to carry out the study.

5) Grants or contracts shall be subject to agreement by the IAC Board co-Chairs and the requesting agency.

6) Administrative and staffing responsibility for the study shall be designated by IAC Board co-Chairs, as deemed most appropriate, either to the IAC Secretariat or, on the basis of a contract with the IAC Secretariat, to an Academy or consortium of Academies.

7) Requests for recommendations for study panel members and reviewers shall be sent electronically to all IAP Academies. Requesting agencies may also suggest names of potential panelists, but the final decision on panel members rests with the IAC. In consultation with IAC Board members (or designated subgroup), appropriate staff shall prepare a slate of panel candidates for two-thirds majority approval by the full

Board, by mail or electronic ballot. The Board may delegate responsibility for approval of panel members to the IAC Board co-Chairs.

8) In cases where candidates for a study panel or other experts have commercial interests or affiliation with political or profit-making organizations directly concerned with the matter of the report, the IAC Board co-Chairs shall ascertain if these individuals can appropriately serve on the study panel for purposes of balance, expertise, and independence. A written statement on the issue, for public release, shall be required of these study panelists where deemed necessary by IAC Board co-Chairs.

9) The Panel shall undertake the work. The IAC Board and the requesting agency shall be kept apprised of panel progress in completing the assignment.

10) All IAC Board members shall be invited to recommend study report reviewers.

11) The reviewers shall be asked to consider the quality of the analysis and fulfillment of study objectives in the draft report. The IAC Board co-Chairs shall appoint two review monitors to determine which reviewer issues shall be addressed by the study panel. The study panel may be asked to consider modifications of the report.

12) The IAC Board co-Chairs shall decide that the report is ready for publication after the review monitors have determined that the study panel has adequately responded to issues raised by the reviewers and the monitors. The IAC Board co-Chairs shall inform the IAC Board whether, in their judgment, the study and review have been conducted satisfactorily in accordance with the IAC Rules of Procedures.

13) The final report shall be publicly issued to requesting agency and to the general public.

Article 4. REPORT DISSEMINATION

1) All final IAC study reports shall be made public with the IAC Board deciding, in each case, on the mechanisms of dissemination.

2) All IAC-published reports shall be issued as authored by members of the study panel, with an explanation that the panel has been selected by the IAC and that the report has satisfactorily completed the IAC review process.

APPENDIX VI: Procedures for the Production and Review of Proactive Academy Reports and Statements of the Royal Academy of Engineering

Excerpts

1. Purpose

The purpose of this document is to outline Quality Assurance procedures designed to ensure the maintenance of the Academy's reputation with respect to its published reports and statements.

2. Scope

These procedures apply to all proactive reports, statements and other documents which are intended for publication by The Royal Academy of Engineering, irrespective of the originator.

The procedures do not apply to responses to Parliament and others, which are dealt with as set out in the (1998) Academy report 'Mechanisms for the provision of advice to external bodies'. Nor do the procedures apply to personal statements which The President might make from time to time in response to national events.

3. Key Elements

It is recognised that reports produced by The Academy will vary considerably in their scope, importance and method of production. However, there are certain features common to all which need to be managed and checked. They are as follows:

- Terms of Reference
- Composition of the working group responsible for producing the report
- Call for evidence
- Circulation of drafts for comment
- Review procedure
- Final approval procedure

These features are to dealt with by following the procedures outlined below:

3.1 Terms of Reference.

Terms of reference define the scope of the study and the deliverables expected.

Every study which is expected to produce a report or statement for publication should begin with the definition of Terms of Reference and their approval by the sponsoring Standing committee (or Council).

It should be understood by all concerned that the Terms of Reference will serve as the primary benchmark against which final review and approval will be conducted.

3.2 Composition of the Working Group responsible for producing a report

The Chairman of the sponsoring standing committee will choose a chairman for the working group in consultation with the secretariat and will invite the appointed chairman to choose additional members in a similar way.

Whenever a working group is formed The Academy will post a notice on its public Web Site which announces the Terms of Reference for the study concerned, giving details of the proposed Chairman and members. Comments or objections will be invited within three weeks. The membership of the working group may subsequently be amended following consultation between the responsible member of the secretariat and the chairman of the sponsoring Standing committee.

At the first meeting of the working group members will be invited to disclose any background, bias or vested interests which are relevant to the study in question. Details will be recorded and summarised in the final report. Members may propose amendments or additions to the working group at this point.

It is recognised that on many issues which are studied by The Academy those who have most to contribute are people with views coloured by experience. It is therefore essential to the acceptability of the final report that the working group is constructed taking into account the nature and sensitivity of the subject matter as well as the need for balance in, and preferably the avoidance of, vested interests. It is also important that the working group contains the competence to address all aspects of its Terms of Reference. For example, economic and social issues associated with an engineering study may require the involvement of appropriate external experts in the working group.

There will be occasions when an issue is very contentious and all Fellows who have experience in the field have strongly vested interests themselves. In consequence the likelihood of achieving balance and a dispassionate consensus is nil. To meet such a challenge an effective solution, which has been employed by The Academy, can be to establish a working group which has no relevant expertise and then to take evidence in a manner similar to a House of Lords committee. This can work very effectively provided appropriate steps are taken to ensure that draft reports are circulated and review is thoroughly undertaken.

It is also important to be aware that on certain issues whilst there may be a consensus among Fellows this may not be shared outside The Academy. In such situations it is important that this is overtly recognised, if not in the composition of the working group, then within the report itself

Underlying these general statements is a challenge to those responsible for establishing any working group, particularly where the ensuing report is intended to influence

Government policy. This is to ensure, as far as is practicable, that readers of the report will not have reason to criticise the working group for presumed bias, vested interest, lack of competence or lack of balance.

3.3 Calls for evidence

If The Academy wishes a report to influence public policy then it is highly desirable that all those outside the Academy who might have views on the subject should be invited to submit evidence. This should be done by posting a notice on the Academy's Web site inviting evidence by a certain date and by the despatch of press releases to those organisations and individuals believed to have something to contribute.

There will be occasions when the working group believes that it need not consult outside the group itself or known experts within the Academy. Whilst this may be acceptable in particular cases, particularly where all that is sought is a consensus opinion among Fellows of The Academy, the first meeting of the group should be invited to discuss this issue and to reach a formally recorded decision.

Written evidence should be supplemented by invited oral evidence on occasion.

3.4 Circulation of drafts for comment

Before a report reaches the review stage it is incumbent on the working group chairman to ensure that the draft is circulated for comment to all those who have contributed evidence. The draft should also be put on the Academy's Web Site inviting any Fellow to comment if they wish.

Whilst it is not expected that everyone will agree with all of the report it is important to pick up errors of fact and understanding which could undermine the conclusions. This consideration outweighs any potential reduction in media interest at the time of the report's publication.

3.5 The Review procedure

When a report is complete in every respect, in the form in which the working group would like it to be published, it shall be subject to review by an independent panel of Fellows prior to being sent to the relevant Standing committee for approval.

The review for a particular report shall be chaired by a member of the standing committee who has had no involvement in the report production process. He will be assisted in the review by not less than two or more than four members of the Academy's panel of reviewers. The chairman of the review panel will decide the appropriate number.

The panel of reviewers will comprise eight Fellows of The Academy who have volunteered their services for three years. They will expect to be called upon from time to time to review reports on any topic, for any standing committee, against criteria similar to those shown in Annex A. The panel will have a range of disciplines within it but expertise in a field is not a prime requirement for a reviewer. Equally, the reviewer is expected to address the report dispassionately and regardless of whether he or she agrees with the conclusions.

Reviewers will remain anonymous and will not be named in the published report.

The report from each reviewer will be sent to the Chairman of the review panel who will pass on the comments, without attribution, to the chairman of the working group. The group must then consciously address each point which is made and record its decisions. Only when the chairman of the review panel is satisfied that the reviewers' comments have been fully and seriously addressed may the report be sent for final approval by the sponsoring standing committee.

3.6 Final approval

Once the report has been reviewed to the satisfaction of the chairman of the review panel he or she will present it to one of the regular meetings of the sponsoring standing committee. The chairman of the working group responsible for the report shall not be present on this occasion, though he or she may have been present to give progress reports on previous occasions.

The standing committee will be invited to give (or withhold) its approval to the report, taking into account the views of the independent reviewers and forming its own judgement about the suitability of the report as an Academy publication.

Where significant changes are requested in a report by a sponsoring standing committee the amended report shall be sent back to the chairman of the committee for final approval.

4. Final note

No set of procedures can cover all possible reports which might be produced by the Academy. However, the underlying principles are clear enough and should be followed when the procedures are found to be wanting or inappropriate.

APPENDIX VII: Study Procedures of the International Council of Academies of Engineering and Technological Sciences

Excerpts

...

The consensus conclusions and recommendations of a CAETS study are the responsibility of the study committee; as a group they are its sole authority or author of record. The role of CAETS is to oversee the proper application of the process for conducting such studies. Member academies may propose appropriate experts for the study committee or other individuals needed in the study process. The preface of the report of any CAETS study will explain this process and will state that CAETS member academies have not endorsed the recommendations or conclusions of the report. These Procedures will be updated to reflect experience.

The Process

1. Any member academy or group of member academies may propose a topic for possible study by CAETS. Such proposals may be made and agreed at Council meetings. When proposed between Council meetings, the Board of Directors may agree or recommend consideration by the Council at its next meeting.

2. Once a topic is agreed for possible study by CAETS, a specific scope of work, including an estimated time schedule and needed resources, is prepared, normally by a group appointed by the CAETS President, including one or more experts nominated by each interested member academy. This group considers any proposed amendments.

3. The completed scope of work, including schedule and resources, is reviewed and approved by the CAETS Executive Committee. Appropriate criteria for approval include:

- importance of the scope of work to the international engineering community;
- uniqueness of CAETS to address the subject;
- existence of sufficient factual information on which to base authoritative recommendations and conclusions;
- a well-defined target audience that could implement the recommendations of the report; and
- likely availability of funding and resources needed for the conduct of the study.

4. Once the scope of work is approved, any necessary funding and other resources may be sought. At the same time member academies are invited to nominate experts to serve on the study committee. After appropriate consultation, including with member academies, the CAETS President appoints the chair and members of

the Study Committee, ensuring expertise appropriate to the study, balanced biases, and no financial conflict of interest. Committee members serve as individuals, not as representatives of their academies or their organizations.

5. With the Study Committee appointed and necessary resources assembled, the Committee begins work. The CAETS President is responsible for oversight of the work of the Committee to ensure the Committee focuses on its scope of work and within its schedule and funds.

6. For the purpose of gathering information relevant to the study, the Committee may hold workshops, or meet with invited experts. Normally, such meetings will be open to the public. However, meetings at which the Committee discusses its recommendations and conclusions normally will not be open to the public.

7. When the Study Committee finishes its draft consensus report, including, if necessary, any dissenting opinions, a Review Group, with expertise similar to that of the Study Committee and selected and appointed in the same manner as the Study Committee, reviews the draft report. The Study Committee responds to each of the comments and suggested changes made by the Review Group. The CAETS President is responsible for refereeing this process to ensure that all Review Group comments are appropriately incorporated.

8. Once the review process is complete, the report is then published and distributed by CAETS. Members of the Study Committee, as well as member academies, may then comment publicly on its recommendations and content.